## Wau Ecology Institute
### Handbook No. 13

# Plants of New Guinea and the Solomon Islands
## Dictionary of the Genera and Families
## of Flowering Plants and Ferns

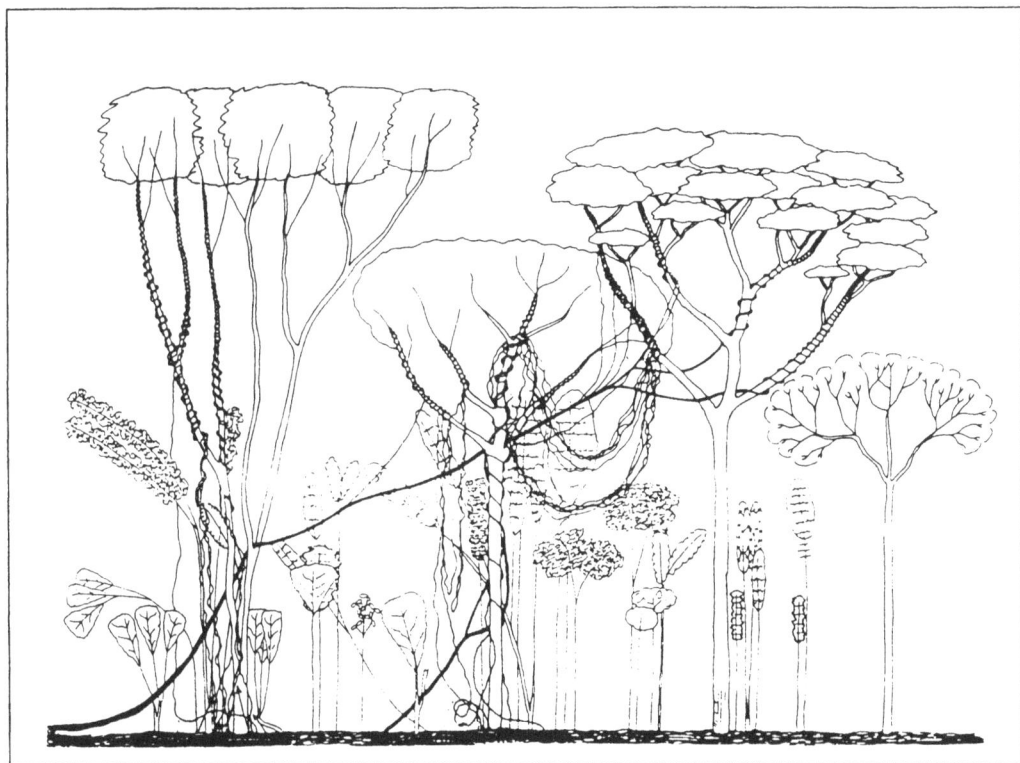

by

Robert Höft

June, 1992

Revision of the Botany Bulletin No. 3 (1973)

# Plants of New Guinea and the Solomon Islands
## Dictionary of the Genera and Families
## of Flowering Plants and Ferns

by

Robert Höft

with a foreword by

Harry Sakulas

**Plants of New Guinea and the Solomon Islands: Dictionary of the Genera and Families of Flowering Plants and Ferns (Wau Ecology Institute Handbook, 13) by Robert Höft**

ISBN 978-9980-945-84-6

First published in 1992 by Wau Ecology Institute
P.O. Box 77
Wau, Morobe Province
Papua New Guinea

Reprinted with permission by the University of Papua New Guinea Press and Bookshop

University of Papua New Guinea Press and Bookshop
P.O. Box 413
University PO, NCD
Papua New Guinea

# Foreword

The botanist working in New Guinea and the Solomon Islands do not have an up dated dictionary of the generic and familiy names of flowering plants and ferns. This revised updating is essentially the important task of the book.

It is significant to note that the updating replaces about a quarter of the old names that were used in the old dictionary (Botany Bulletin No. 3, March, 1973). The Island so rich and diverse in plant live where new genus or even families are expected to be found deserves regular updating.

It is not surprising that endemism is quite high comparatively. With the production of this book, we believe the curators of botanical collections will find it a useful tool to relabell the specimens with the new name. Similarly, the curators of botanical gardens and arboretum are expected to do the same thing.

However, it must be admitted that the debate on change of names and reclassification of families and genus is always expected to remain unsolved among the systematic biologist for some time. It is important for the users to be mindful of the possible change.

My sincere gratitude to the author, Robert Hoeft for compiling the revised dictionary quickly and promptly. I recommend every botanist working in the region to obtain a copy as soon as possible.

Harry Sakulas
Director
Wau Ecology Institute

# Contents

# Introduction

This dictionary has been prepared primarily for professional botanists and for the students of biology and forestry in P.N.G. It is based on the latest edition of the "Dictionary of the Generic and Family names of Flowering plants and Ferns of the New Guinea and South-West Pacific Region" (Botany Bulletin No. 3, March 1973). It includes the genera found in Papua New Guinea with the islands, in West Papua (Irian Jaya) and in the Solomon islands. Also included are introduced plant genera and those of economical importance.

The generic names have been matched with "The plant-book" by D.J. Mabberley (Cambridge University Press, 1987) and the "Checklist of generic names in Malesian botany" by C.G.G.J. van Steenis (Flora Malesiana Foundation, Leiden, 1987). Some older names have been re-checked in the "Dictionary of the Flowering Plants & Ferns" by J.C. Willis (7th edition, Cambridge University Press, 1966) and the various volumes of the Flora Malesiana to give the correct synonymies. A number of recent revisions of single genera of flowering plants have been consulted. The bulk of information has been published in Blumea, in the Australian Journal of Botany, in Reinwardtia, in the Journal of the Arnold Arboretum, in Science of New Guinea and by the Rijksherbarium Leiden and the Royal Botanic Gardens, Kew . For ferns the "Key to the Families and Genera of Pteridophytes in Papuasia" by J.R. Croft (1975) has been used as an additional source of information.

During the preparation of this booklet it appeared that a large number of names of genera reported from New Guinea and the Solomon Islands are no longer valid or actually do not occur in the area. In order to facilitate the updating of the nomenclature the valid generic names are printed in bold letters and the authority is given. In single cases, however, where changes have not yet been generally accepted the old names were retained, e.g. Euodia. Collections from outside the area which have crept into the literature are kept in normal printing and bear a remark "no record." Generic names in brackets indicate that the name of the entry is sometimes included in the genus or different unaccepted spellings.

The classification of ferns follows the system introduced by J.A. Crabbe, A.C. Jermy and J.T. Mickel ("New generic sequence for the pteridophyte herbarium," Fern Gazette 11: 141-162, 1975) which is used at the National Herbarium in Lae. Each genus is presented with its code number.

The classification of the families of flowering plants adapted here follows largely A. Cronquist's "Integrated system of classification of flowering plants" (Columbia University Press, 1981). However, a few modifications are made:

- the Leguminosae are split in three families of their own (Caesalpiniaceae, Fabaceae, Mimosaceae)
- Lobeliaceae are excluded from Campanulaceae
- Butomaceae are included in Limnocharitaceae
- Stilaginaceae are separated from Euphorbiaceae and Avicenniaceae are separated from Verbenaceae
- Amaryllidaceae and Hypoxidaceae are separated from Liliaceae
- Sansevieraceae and Dracaenaceae are separated from Agavaceae.

The families and genera within the families are arranged in alphabetical order. Subfamilies are given only for pteridophytes.

For each record the number of species reported for the Solomon Islands (from I.R. Hancock & C.P. Anderson, 1988, Flora of the Solomon Islands, Research Bull. No. 7, Dodo Creek Research Station), for New Guinea (incl. Solomons) and the number worldwide is given. Similarly, for each family the number of genera and species in the area are compared to the overall number of genera and species. For example:

## ACANTHACEAE                    32(129) / 357(4350)
Acanthus L.                          2 / 4 / 30

Read: There are 32 genera with 129 species in the Acanthaceae family which occur in New Guinea and the Solomon Islands. Worldwide the Acanthaceae comprise 357 genera with 4,350 species. Within the genus Acanthus two species occur in the Solomon Islands, 4 in the whole region and 30 are reported worldwide.

However, the accuracy of these figures may vary considerably. Where not available from publications the species number was obtained from the card file system at the National Herbarium. In some cases synonyms may have been counted additionally, in others the records may have been missing. According to these sources the total number of species for New Guinea and the Solomon Islands amounts to 15,230.

Genera endemic in New Guinea or in Papuasia are marked. A number of formerly endemic genera has been included in genera of wider distribution. Some of them are given below.

| | | |
|---|---|---|
| Abromeitia Mez. | now: Fittingia | Myrsinaceae |
| Ancistragrostis S.T. Blake | now: Deyeuxia | Poaceae |
| Ancylacanthus Lindau | now: Ptyssiglottis | Acanthaceae |
| Anthobembix Perkins | now: Steganthera | Monimiaceae |
| Bamlera Schumann & Lauterb. | now: Astronidium | Melastomataceae |
| Cephalohibiscus Ulbr. | now: Thespesia | Malvaceae |
| Coombea P. Royen | now: Medicosma | Rutaceae |
| Echinocitrus Tanaka | now: Triphasia | Rutaceae |
| Guillainea Vieill. | now: Alpinia | Zingiberaceae |
| Lamechites Markgraf | now: Micrechtites | Apocynaceae |
| Leptosiphonium F. Muell. | now: Ruellia | Acanthaceae |
| Myrmedoma Becc. | now: Myrmephytum | Rubiaceae |
| Nothoruellia Bremek. & Nannenga-Bremek. | now: Ruellia | Acanthaceae |
| Oxychlamys Schltr. | now: Aeschynanthus | Gesneriaceae |
| Papuzilla Ridley | now: Lepidium | Brassicaceae |
| Phyllapophysis Mansf. | now: Catanthera | Melastomataceae |
| Scrobicularia Mansf. | now: Poikilogyne | Melastomataceae |
| Tripetalum Schumann | now: Garcinia | Clusiaceae |
| Xerocarpa H.J. Lam | now: Teijsmanniodendron | Verbenaceae |

The number of endemic genera in the region amounts to 97 (in Papuasia), 84 of which are endemic in New Guinea. This is equivalent to a rate of endemism of about 5 percent on the generic level.

As the data presented here will always be subject to changes and specialists on single groups are likely to have more accurate informations the data set is available from the author both as a 'dbase' and an ASCII-file to allow an easy updating. Send your diskette to Wau Ecology Institute, P.O. Box 77, Wau, Morobe Province, Papua New Guinea.

The following abbreviations and symbols are used:

---

coll. uncertain ..... the entry is based on a single specimen whose origin is uncertain

cult. ................ the genus or most of its species are cultivated

econ. ............... the genus or some species have economic value

ed. ................. the genus or some species are edible

ident.? ............. the entry is based on a single specimen whose identification is uncertain

intro. ............... the genus or most of its species are introduced

intro.? .............. the genus is probably introduced

nat. ................ the genus or most of its species are introduced and became naturalized

orn. ................ the genus or most of its species are grown as ornamentals

Printing:

bold face .......... generic name accepted for the area

regular type ........ genus not accepted for the area or genus reported to occur but no material was seen at the National Herbarium

---

## Acknowledgements

I wish to express my sincere gratitude to Matthew Jebb who provided a huge amount of valuable informations and made excellent suggestions.

Special thanks are also due to David Frodin who was prepared to share his incredible knowledge of the flora of the region.

Furthermore, I would like to thank the staff of the National Herbarium, Lae, for their support in collecting the informations. In particular, I wish to thank Robert Kiapranis, Osia Gideon, Neville Howcraft and Paul Katik.

I am grateful to Ollo Gebia and Peter Amatus for their assistance in compiling the list of vernacular names and to Larry Orsak for his comments and suggestions.

Finally I would like to thank Harry Sakulas for his support and interest during the ongoing of the project.

# I. List of the Genera in New Guinea and the Solomon Islands

Abarema = Archidendron .............................................. Mimosaceae

**Abelmoschus** Medik. (Hibiscus) ..................................... Malvaceae

**Abrodictyum** C. Presl .............................................. Hymenophyllaceae

Abroma = Ambroma.................................................. Sterculiaceae

**Abromeitia** Mez. = Fittingia ...................................... Myrsinaceae

**Abrotanella** Cass.................................................. Asteraceae

**Abrus** Adans........................................................ Fabaceae

**Abutilon** Miller ................................................... Malvaceae

**Acacia** Miller ..................................................... Mimosaceae

**Acaena** Mutis ex L. ................................................ Rosaceae

**Acalypha** L......................................................... Euphorbiaceae

**Acanthophippium** Blume ............................................ Orchidaceae

**Acanthus** L........................................................ Acanthaceae

**Aceratium** DC. ..................................................... Elaeocarpaceae

Achasma = Amomum................................................ Zingiberaceae

**Achillea** L......................................................... Asteraceae

Achras = Manilkara ................................................ Sapotaceae

**Achyranthes** L..................................................... Amaranthaceae

Achyrospermum Blume ................................................ Lamiaceae .................... no record

**Acianthus** R. Br................................................... Orchidaceae

Ackama = Caldcluvia................................................ Cunoniaceae

Aclisia = Pollia .................................................... Commelinaceae

Acmena = Syzygium................................................ Myrtaceae

**Acorus** L.......................................................... Araceae, possibly separate family

**Acriopsis** Blume .................................................. Orchidaceae

**Acrocephalus** Benth................................................ Lamiaceae

Acroceras Stapf (Panicum) .......................................... Poaceae ..................... no record

**Acronychia** Forster & G. Forster.................................. Rutaceae

**Acrophorus** C. Presl. ............................................. Aspleniaceae (Dryopt.)

**Acrosorus** Copel. ................................................. Grammitidaceae

**Acrostichum** L. ................................................... Adiantaceae (Pter.)

**Acsmithia** Hoogl. ................................................. Cunoniaceae

**Actephila** Blume .................................................. Euphorbiaceae

**Actinodaphne** Nees................................................. Lauraceae

Actinophloeus = Ptychosperma ....................................... Arecaceae

**Actinorhytis** H. Wendl. & Drude .................................. Arecaceae

Actoplanes = Donax ................................................. Marantaceae

Adelia see Mallotus, Spatiostemon .................................. Euphorbiaceae

Adelonenga = Hydriastele ........................................... Arecaceae

**Adenanthera** L. ................................................... Mimosaceae

**Adenia** Forssk. ................................................... Passifloraceae

**Adenium** Roemer & Schultes ....................................... Apocynaceae

**Adenoncos** Blume .................................................. Orchidaceae

Adenosacme = Mycetia ............................................... Rubiaceae

**Adenosma** R. Br. ....................................................... Scrophulariaceae

**Adenostemma** Forster & G. Forster ............................. Asteraceae

**Adiantum** L. ............................................................... Adiantaceae (Adiant.)

Adina see Metadina ..................................................... Rubiaceae

**Adinandra** Jack ........................................................... Theaceae

Adisca = Mallotus ....................................................... Euphorbiaceae

**Aegialitis** R. Br. .......................................................... Plumbaginaceae

**Aegiceras** Gaertner ..................................................... Myrsinaceae

**Aeginetia** L. ................................................................ Orobanchaceae

**Aegopogon** Humb. & Bonpl. ....................................... Poaceae

**Aerides** Lour. .............................................................. Orchidaceae

**Aerva** Forssk. ............................................................. Amaranthaceae

**Aeschynanthus** Jack ................................................... Gesneriaceae

**Aeschynomene** L. ........................................................ Fabaceae

**Afgekia** Craib ............................................................. Fabaceae

Afzelia see Intsia ......................................................... Caesalpiniaceae

Agalmyla Blume ........................................................... Gesneriaceae ................. no record

**Aganope** Miq. ............................................................. Fabaceae

**Agapetes** D. Don ex G. Don. f. ................................... Ericaceae

**Agatea** A. Gray ........................................................... Violaceae

**Agathis** Salisb. ........................................................... Araucariaceae

**Agave** L. .................................................................... Agavaceae

**Ageratum** L. ............................................................... Asteraceae

**Aglaia** Lour. ............................................................... Meliaceae

Aglaiopsis = Aglaia ..................................................... Meliaceae

**Aglaomorpha** Schott ................................................... Polypodiaceae (Dryn.)

**Aglaonema** Schott ...................................................... Araceae

**Aglossorhyncha** Schltr. ................................................ Orchidaceae

**Agonis** (DC.) Sweet (Sinoga) ...................................... Myrtaceae

Agropyron see Brachypodon .......................................... Poaceae

**Agrostis** L. ................................................................. Poaceae

**Agrostistachys** Dalz. .................................................. Euphorbiaceae

**Agrostophyllum** Blume ............................................... Orchidaceae

**Ailanthus** Desf. .......................................................... Simaroubaceae

**Aira** L. ...................................................................... Poaceae

**Airosperma** Schumann & Lauterb. ............................... Rubiaceae

**Aistopetalum** Schltr. ................................................... Cunoniaceae

**Ajuga** L. .................................................................... Lamiaceae

**Alangium** Lam. ........................................................... Alangiaceae

**Albertisia** Becc. .......................................................... Menispermaceae

Albertisiella = Pouteria ................................................. Sapotaceae

**Albizia** Durazz. ........................................................... Mimosaceae

**Alchornea** Sw. ............................................................ Euphorbiaceae

Alcinaeanthus = Neoscortechina ..................................... Euphorbiaceae

Aldrovanda L. ...................................................... Droseraceae ................ no record

Alectryon Gaertner ................................................ Sapindaceae

Aleurites Forster & G. Forster .................................... Euphorbiaceae

Allamanda L. ...................................................... Apocynaceae

Allium L. ......................................................... Liliaceae

Allomorphia = Oxyspora ............................................ Melastomataceae

Allophylus L. ..................................................... Sapindaceae

Alloteropsis J. Presl ............................................. Poaceae

Allowoodsonia Markgraf ............................................ Apocynaceae

Alocasia (Schott) G. Don. f. ...................................... Araceae

Alphandia Baillon ................................................. Euphorbiaceae

Alphitonia Reisseck ex Endl. ...................................... Rhamnaceae

Alphonsea Hook. f. & Thomson ...................................... Annonaceae

Alpinia Roxb. ..................................................... Zingiberaceae

Alseodaphne Nees .................................................. Lauraceae

Alsodeia = Rinorea ................................................ Violaceae

Alsomitra (Blume) M. Roemer ....................................... Cucurbitaceae

Alsophila = Cyathea ............................................... Cyatheaceae

Alstonia R. Br. ................................................... Apocynaceae

Alternathera Forssk. .............................................. Amaranthaceae

Althoffia = Trichospermum ......................................... Tiliaceae

Alysicarpus Desv. ................................................. Fabaceae

Alyxia R. Br. ..................................................... Apocynaceae

Amaracarpus Blume ................................................. Rubiaceae

Amaranthus L. ..................................................... Amaranthaceae

Amauropelta Kunze ................................................. Thelypteridaceae

Amblyanthus A. DC. (Conandrium) ................................... Myrsinaceae

Ambroma L. f. ..................................................... Sterculiaceae

Amherstia Wallich ................................................. Caesalpiniaceae

Amischotolype Hassk. .............................................. Commelinaceae

Ammannia L. ....................................................... Lythraceae

Amomum Roxb. ...................................................... Zingiberaceae

Amoora = Aglaia ................................................... Meliaceae

Amorphophallus Blume ex Decne. .................................... Araceae

Ampelocissus Planchon ............................................. Vitaceae

Ampelopteris Kunze ................................................ Thelypteridaceae

Amphicarpaea Elliott ex Nutt. ..................................... Fabaceae ................ no record

Amphineuron Holttum ............................................... Thelypteridaceae

Amphipterum (Copel.) Copel. (Mecodium) ............................ Hymenophyllaceae

Amydrium Schott ................................................... Araceae

Amyema Tieghem .................................................... Loranthaceae

Amylotheca Tieghem ................................................ Loranthaceae

Anacardium L. ..................................................... Anacardiaceae

**Anacolosa** (Blume) Blume ............................................ Olacaceae
**Anakasia** Philipson ..................................................... Araliaceae
**Anamirta** Colebr. ........................................................ Menispermaceae
**Ananas** Miller ........................................................... Bromeliaceae
**Anaphalis** DC. ........................................................... Asteraceae
**Anarthropteris** Copel. .................................................. Grammitidaceae
Ancistragrostis = Calamagrostis .................................... Poaceae
Ancistrocladus Wallich .................................................. Ancistrocladaceae .......... no record
Ancistrocladus see Durandea ......................................... Linaceae
Ancistrum = Acaena ..................................................... Rosaceae
Ancylacanthus Lindau = Ptyssiglottis ............................ Acanthaceae
**Andredera** Juss. .......................................................... Basellaceae
Androcephalium = Lunasia ............................................. Rutaceae
**Andropogon** L. ........................................................... Poaceae
**Androsace** L. (Primula) ............................................... Primulaceae
Andruris = Sciaphila ..................................................... Triuridaceae
**Aneilema** R. Br. .......................................................... Commelinaceae
Angelesia = Licania ...................................................... Chrysobalanaceae
**Angelonia** Bonpl. ........................................................ Scrophulariaceae
**Angiopteris** Hoffm. ..................................................... Marattiaceae
**Angraecopsis** Kraenzlin ............................................... Orchidaceae
**Angraecum** Bory ......................................................... Orchidaceae
**Aniseia** Choisy ........................................................... Convolvulaceae
Aniselytron Merr. ......................................................... Poaceae ..................... no record
Anisocampium C. Presl. ................................................ Aspleniaceae (Athyr.) ..... no record
Anisogonium = Diplazium ............................................. Aspleniaceae (Athyr.)
**Anisomeles** R. Br. ....................................................... Lamiaceae
**Anisoptera** Korth. ....................................................... Dipterocarpaceae
**Annesijoa** Pax & K. Hoffm. ......................................... Euphorbiaceae
**Annona** L. ................................................................. Annonaceae
**Anodendron** A. DC. ..................................................... Apocynaceae
**Anoectochilus** Blume ................................................... Orchidaceae
**Anogramma** Link ........................................................ Adiantaceae (Adiant.)
Anomopanax = Mackinlaya ............................................ Araliaceae
Anoniodes see Sloanea .................................................. Elaeocarpaceae
**Anotis** DC. (Neanotis) .................................................. Rubiaceae
Anplectrum = Diplectria ............................................... Melastomataceae
Anthistiria = Themeda .................................................... Poaceae
Anthobembix Perkins = Steganthera ............................... Monimiaceae
**Anthobolus** R. Br. (Exocarpos) ..................................... Santalaceae
**Anthocarapa** Pierre ..................................................... Meliaceae
**Anthocephalus** A. Rich. ................................................ Rubiaceae
Antholoma see Sloanea .................................................. Elaeocarpaceae

**Anthorrhiza** Huxley & Jebb ............................... Rubiaceae
**Anthoxanthum** L. .................................................. Poaceae
Anthurium see Pothos ............................................. Araceae
**Antiaris** Leschen. .................................................. Moraceae
**Antiaropsis** Schumann ........................................... Moraceae
**Antidesma** L. ....................................................... Stilaginaceae
**Antigonon** Endl. .................................................. Polygonaceae
Antirhea see Guettardella ....................................... Rubiaceae
**Antrophyum** Kaulf. ............................................... Adiantaceae (Vitt.)
**Aphanamixis** Blume.............................................. Meliaceae
**Aphananthe** Planchon............................................ Ulmaceae
Aphania = Lepisanthes ............................................ Sapindaceae
**Aphanomyrtus** = Syzygium ..................................... Myrtaceae
**Aphelandra** R. Br. ................................................ Acanthaceae
Aphoma = Iphigenia ............................................... Liliaceae
**Aphyllorchis** Blume .............................................. Orchidaceae
**Apium** L. ............................................................ Apiaceae
**Apluda** L. ........................................................... Poaceae
**Aponogeton** L. f. .................................................. Aponogetonaceae
Aporetica = Allophylus ............................................ Sapindaceae
Aporuellia see Ruellia .............................................. Acanthaceae
Aporum see Dendrobium ........................................... Orchidaceae
**Aporusa** Blume (Aporosa) ...................................... Euphorbiaceae
**Apostasia** Blume .................................................. Orchidaceae
**Appendicula** Blume ............................................... Orchidaceae
**Aquilaria** Lam..................................................... Thymelaeaceae
**Arachis** L. .......................................................... Fabaceae
**Arachniodes** Blume............................................... Aspleniaceae (Dryopt.)
**Arachnis** Blume ................................................... Orchidaceae
Araiostegia Copel. .................................................. Davalliaceae (Davall.) .... no record
**Aralia** L............................................................. Araliaceae
**Araucaria** Juss. ................................................... Araucariaceae
**Arcangelisia** Becc. ................................................ Menispermaceae
**Archboldia** E. Beer & H.J. Lam ............................... Verbenaceae
**Archboldiodendron** Kobuski ................................... Theaceae
**Archidendron** F. Muell........................................... Mimosaceae
Archontophoenix H.A. Wendl. & Drude ...................... Arecaceae ................... no record
Arcypteris = Pleocnemia .......................................... Aspleniaceae (Tect.)
**Ardisia** Sw.......................................................... Myrsinaceae
**Areca** L. ............................................................ Arecaceae
**Arenga** Lab......................................................... Arecaceae
**Argemone** L........................................................ Papveraceae
**Argostemma** Wallich ............................................. Rubiaceae
**Argusia** Boehmer (Tournefortia) .............................. Boraginaceae

Argyrocalymma = Carpodetus .......................................... Grossulariaceae
Argyrodendron F. Muell. (Heritiera) ............................... Sterculiaceae ............... no record
Arisacontis see Cyrtosperma .......................................... Araceae
Aristida L. ................................................................. Poaceae
Aristolochia L. ........................................................... Aristolochiaceae
Aristotelia L'Hérit. (Sericolea) ...................................... Elaeocarpaceae
Aromadendron = Magnolia ............................................ Magnoliaceae
Arrhenechthites Mattf. ................................................. Asteraceae
Artabotrys R. Br. ........................................................ Annonaceae
Artanema D. Don ........................................................ Scrophulariaceae
Artemisia L. ............................................................... Asteraceae
Arthraxon Pal. ............................................................ Poaceae
Arthrobothrya J. Sm. ................................................... Aspleniaceae (Lom.)
Arthrochilus F. Muell. (Spiculaea) ................................. Orchidaceae
Arthrocnemon Moq. (Arthrocnemum) see Halosarcia .......... Chenopodiaceae
Arthromeris (T. Moore) J. Sm. ...................................... Polypodiaceae (Micro.)... no record
Arthrophyllum Blume .................................................. Araliaceae
Arthropodium R. Br. .................................................... Liliaceae
Arthropteris J. Sm. ex Hook. f. ..................................... Davalliaceae (Oleand.)
Artocarpus Foster & Foster f. ........................................ Moraceae
Arundinaria Michaux .................................................... Poaceae
Arundinella Raddi ....................................................... Poaceae
Arundo L. .................................................................. Poaceae
Arytera Blume ............................................................ Sapindaceae
Asarina Miller ............................................................ Scrophulariaceae .......... no record
Ascarina Foster & Foster f. .......................................... Chloranthaceae
Asclepias L. ............................................................... Asclepiadaceae
Ascocentrum Schltr. ex J.J. Sm. .................................... Orchidaceae ................ no record
Ascoglossum Schltr. (Sarcochilus) .................................. Orchidaceae
Asparagus L. .............................................................. Liliaceae
Aspidium = Tectaria .................................................... Aspleniaceae (Tect.)
Aspidocarya = Cardiopteris ........................................... Cardiopteridaceae
Asplenium L. .............................................................. Aspleniaceae (Asplen.)
Astelia Banks & Sol. ex R. Br. ...................................... Liliaceae
Astelma R. Br. see Helichrysum ..................................... Asteraceae
Astelma Schl. = Papuastelma ........................................ Asclepiadaceae
Aster L. .................................................................... Asteraceae
Asteromyrtus = Melaleuca ............................................ Myrtaceae
Astilbe Buch.-Ham. ex D. Don ...................................... Saxifragaceae
Astrebla F. Muell. ....................................................... Poaceae ..................... no record
Astronia Blume ........................................................... Melastomataceae
Astronidium A. Gray .................................................... Melastomataceae
Asystasia Blume .......................................................... Acanthaceae

# I. List of the Genera in New Guinea and the Solomon Islands

Atalantia Corr. Serr....................................................Rutaceae

Atalaya Blume .........................................................Sapindaceae

Athyrium Roth .......................................................Aspleniaceae (Athyr.)

Atylosia Wight & Arn. .............................................Fabaceae

Aucoumea Pierre ....................................................Burseraceae

Aulacolepis = Aniselytron ........................................Poaceae

Aulostylis = Calanthe...............................................Orchidaceae

Austrobaileya C. White ...........................................Austrobaileyaceae

Austrobuxus see Kairothamnus ..................................Euphorbiaceae

Averrhoa L..............................................................Oxalidaceae

Avicennia L. ...........................................................Avicenniaceae

Axonopus see Alloteropsis ........................................Poaceae

Azadirachta A. Juss. ................................................Meliaceae

Azima Lam. ...........................................................Salvadoraceae

Azolla Lam. ...........................................................Azollaceae

Baccaurea Lour. .....................................................Euphorbiaceae

Baccharis see Pluchea ..............................................Asteraceae

Backhousia Hook. & Harvey ....................................Myrtaceae

Bacopa Aublet ........................................................Scrophulariaceae

Bacularia = Linospadix.............................................Arecaceae

Badusa A. Gray ......................................................Rubiaceae

Baeckea L. .............................................................Myrtaceae

Baileyoxylon C. White..............................................Flacourtiaceae

Balanophora Forster & G. Forster...............................Balanophoraceae

Balanops Baillon......................................................Balanopaceae ...............no record

Balantium = Dicksonia .............................................Cyatheaceae

Ballota see Hyptis ...................................................Lamiaceae

Baloghia see Fonatainea ...........................................Euphorbiaceae

Bambusa Schreber....................................................Poaceae

Bamlera Schumann & Lauterb. = Astronidium .................Melastomataceae

Bania = Carronia ....................................................Menispermaceae

Banksia L. f. ..........................................................Proteaceae

Barclaya = Hydrostemma ..........................................Nymphaeaceae

Barkerwebbia = Heterospathe.....................................Arecaceae

Barleria L...............................................................Acanthaceae

Barringtonia Forster & G. Forster ..............................Barringtoniaceae

Barrotia = Pandanus ...............................................Pandanaceae

Basella L.................................................................Basellaceae

Basilicum Moench ...................................................Lamiaceae

Basisperma C. White ...............................................Myrtaceae

Bassia = Madhuca....................................................Sapotaceae

Bassia All. ..............................................................Chenopodiaceae...........no record

Batis P. Browne ......................................................Bataceae

Bauerella Borzi .......................................................Rutaceae

Bauhinia L. .............................................. Caesalpiniaceae

Baumea see Machaerina .............................. Cyperaceae

Beauvisagea = Pouteria ............................... Sapotaceae

Beccarianthus see Astronidium .................... Melastomataceae

Beccariella = Pichonia ................................ Sapotaceae

Beccariodendron see Goniothalamus ............. Annonaceae

Begonia L. .............................................. Begoniaceae

Beilschmiedia Nees ................................... Lauraceae

Belliolum = Zygogynum ............................. Winteraceae

Beloperone = Justicia ................................. Acanthaceae

Belvisia Mirbel ....................................... Polypodiaceae (Pleo.)

Benincasa Savi ....................................... Cucurbitaceae

Bennettia = Bennettiodendron ..................... Flacourtiaceae

Bennettiodendron Merr. ............................. Flacourtiaceae

Berchemia Necker ex DC. .......................... Rhamnaceae

Bergia L. ............................................... Elatinaceae

Berrya Roxb. .......................................... Tiliaceae

Betchea = Caldcluvia ................................ Cunoniaceae

Bhesa Buch.-Ham. ex Arn. ......................... Celastraceae

Bidens L. ............................................... Asteraceae

Bignonia see Dolichandrone ........................ Bignoniaceae

Bikkia Reinw. ......................................... Rubiaceae

Biophytum DC. ........................................ Oxalidaceae

Bischofia Blume ...................................... Euphorbiaceae ............. no record

Bixa L. .................................................. Bixaceae

Bleasdalea see Gevuina .............................. Proteaceae

Blechnum L. ........................................... Blechnaceae

Bleekeria = Ochrosia ................................. Apocynaceae

Bletia Ruíz & Pavón ................................. Orchidaceae

Blumea DC. ............................................ Asteraceae

Blumeodendron Kurz ................................. Euphorbiaceae

Blyxa Noronha ex Thouars .......................... Hydrocharitaceae

Boea Comm. ex Lam. ................................ Gesneriaceae

Boehmeria Jacq. ...................................... Urticaceae

Boerhavia L. ........................................... Nyctaginaceae

Boerlagiodendron = Osmoxylon .................... Araliaceae

Boesenbergia Kuntze ................................. Zingiberaceae

Bogoria J.J. Sm. ...................................... Orchidaceae

Bolbitis Schott ........................................ Aspleniaceae (Lom.)

Bombacopsis = Pachira ............................... Bombacaceae

Bombax L. .............................................. Bombacaceae

Bonamia Thouars ..................................... Convolvulaceae

Bonnaya see Lindernia ............................... Scrophulariaceae

Borassus L. ..................................................... Arecaceae
Borreria see Spermacoce .............................. Rubiaceae
Bothriochloa Kuntze (Dichanthium) ............................ Poaceae
Botrychium Sw.................................................... Ophioglossaceae
Bouchardatia Baillon (Melicope)..................... Rutaceae
Bougainvillea Comm. ex Juss. ....................... Nyctaginaceae
Brachiaria (Trin.) Griseb. ............................... Poaceae
Brachionostylum Mattf.................................. Asteraceae
Brachistus Miers........................................... Solanaceae
Brachyachne (Benth.) Stapf............................ Poaceae ...................... no record
Brachychiton Schott & Endl........................... Sterculiaceae
Brachycome Cass........................................... Asteraceae
Brachypodium Pal........................................ Poaceae
Brachystelma R. Br. (Microstemma) ........................... Asclepiadaceae
Brachythalamus = Gyrinops ....................... Thymelaeaceae
Brackenridgea A. Gray ................................ Ochnaceae
Brainea J. Sm. ............................................... Blechnaceae................ no record
Brassaia = Schefflera .................................. Araliaceae
Brassiantha A.C. Sm. .................................. Celastraceae
Brassica L. ................................................... Brassicaceae
Brassiodendron = Endiandra ....................... Lauraceae
Brassiophoenix Burret ................................ Arecaceae
Bredemeyera Willd. (Polygala)..................... Polygalaceae
Breynia Forster & Forster f. ....................... Euphorbiaceae
Bridelia Willd. ............................................. Euphorbiaceae
Briza L. ....................................................... Poaceae
Bromheadia Lindley .................................... Orchidaceae
Bromus L...................................................... Poaceae
Broussonetia L'Hérit. ex Vent. ..................... Moraceae
Browallia L................................................... Solanaceae
Brownea Jacq. .............................................. Caesalpiniaceae
Brownlowia Roxb.......................................... Tiliaceae
Brucea J.F. Miller ......................................... Simaroubaceae
Brugmansia Pers. (Datura) ........................... Solanaceae ................. no record
Bruguiera Lam.............................................. Rhizophoraceae
Bruinsmia Boerl. & Koord. .......................... Styracaceae
Bryantea = Neolitsea .................................. Lauraceae
Bryanthia = Pandanus ................................ Pandanaceae
Bryonia see Melothria .................................. Cucurbitaceae
Bryonopsis = Kedrostis ................................ Cucurbitaceae
Bryophyllum = Kalanchoe ........................... Crassulaceae
Bubbia = Zygogynum ................................ Winteraceae
Buchanania Sprengel...................................... Anacardiaceae
Buchnera L. ................................................. Scrophulariaceae

Buddleia = Buddleja .................................................. Loganiaceae
**Buddleja** L. ............................................................. Loganiaceae
**Buergersiochloa** Pilger ........................................... Poaceae
**Bulbophyllum** Thouars ........................................... Orchidaceae
**Bulbostylis** Kunth (Fimbristylis) ............................ Cyperaceae
**Burckella** Pierre...................................................... Sapotaceae
Bureavella = Pouteria................................................ Sapotaceae
**Burmannia** L. .......................................................... Burmanniaceae
Bursera see Protium................................................... Burseraceae
**Butea** Roxb. ex Willd. ............................................. Fabaceae
**Butomopsis** Kunth (Tenagocharis)........................... Limnocharitaceae
**Byblis** Salisb. .......................................................... Byblidaceae
Bysteropogon see Hyptis ........................................... Lamiaceae
**Cabomba** Aublet...................................................... Cabombaceae
**Cadaba** Forssk. ....................................................... Capparaceae
**Cadetia** Gaudich. .................................................... Orchidaceae
Caelospermum = Coelospermum ................................ Rubiaceae
**Caesalpinia** L.......................................................... Caesalpiniaceae
**Caesia** R. Br. .......................................................... Liliaceae
**Cajanus** DC............................................................. Fabaceae
Calacinum see Muehlenbeckia ................................... Polygonaceae
**Caladium** Vent. ....................................................... Araceae
**Calamagrostis** Adans. (Deyeuxia)........................... Poaceae
Calamintha see Satureja ............................................ Lamiaceae
**Calamus** L. ............................................................. Arecaceae
**Calanthe** R. Br. ....................................................... Orchidaceae
**Caldcluvia** D. Don .................................................. Cunoniaceae
**Caldesia** Parl. (Alisma) .......................................... Alismataceae
**Calliandra** Benth..................................................... Mimosaceae
**Callicarpa** L. ........................................................... Verbenaceae
**Callipteris** Bory (Diplazium) .................................. Aspleniaceae (Athyr.)
Callista = Dendrobium ............................................... Orchidaceae
**Callistopteris** Copel. (Trichomanes) ....................... Hymenophyllaceae
**Callitriche** L............................................................ Callitrichaceae
**Callitris** Vent. ......................................................... Cupressaceae
**Calochilus** R. Br...................................................... Orchidaceae
Calodracon = Cordyline............................................. Agavaceae
**Calogyne** R. Br. ...................................................... Goodeniaceae
Calonyction = Ipomoea............................................... Convolvulaceae
Calophanoides = Justicia............................................ Acanthaceae
**Calophyllum** L. ....................................................... Clusiaceae
**Calopogonium** Desv. ............................................... Fabaceae
**Calotropis** R. Br. .................................................... Asclepiadaceae

Calpidia see Ceodes....................................................Nyctaginaceae

Calpidochlamys see Trophis.........................................Moraceae

**Calycacanthus** Schumann ...........................................Acanthaceae

**Calycosia** A. Gray ....................................................Rubiaceae

**Calymmanthera** Schltr. .............................................Orchidaceae

**Calymmodon** C. Presl. ..............................................Grammitidaceae

Calyptranthus = Syzygium..........................................Myrtaceae

**Calyptrocalyx** Blume ...............................................Arecaceae

Calysaccion see Mammea...........................................Clusiaceae

**Calystegia** R. Br. ...................................................Convolvulaceae ........... no record

Camarotis = Micropera..............................................Orchidaceae

Campanumoea = Codonopsis.......................................Campanulaceae

Campium = Bolbitis ...................................................Aspleniaceae (Lom.)

**Campnosperma** Thwaites ..........................................Anacardiaceae

Campsis Lour. .........................................................Bignoniaceae................ no record

**Camptostemon** Masters ............................................Bombacaceae

**Cananga** (DC.) Hook. f. & Thomson ............................Annonaceae

**Canarium** L............................................................Burseraceae

**Canavalia** DC. ......................................................Fabaceae

Candollea = Stylidium................................................Stylidiaceae

**Canna** L. .............................................................Cannaceae

**Cannabis** L...........................................................Cannabidaceae

**Cansjera** Juss........................................................Opiliaceae

**Canthium** Lam. (Plectronia).....................................Rubiaceae

**Capillipedium** Stapf (Dichanthium) ............................Poaceae

Capitularia = Chorizandra .........................................Cyperaceae

Capitularina = Chorizandra.........................................Cyperaceae

**Capparis** L...........................................................Capparaceae

Capraria see Lindernia ..............................................Scrophulariaceae

**Capsella** Medikus...................................................Brassicaceae

**Capsicum** L..........................................................Solanaceae

**Carallia** Roxb........................................................Rhizophoraceae

**Carapa** Aublet (Xylocarpus)......................................Meliaceae

**Cardamine** L. .......................................................Brassicaceae

Cardiocarpus = Soulamea ..........................................Simaroubaceae

Cardiophora = Soulamea.............................................Simaroubaceae

**Cardiopteris** Wallich ex Royle....................................Cardiopteridaceae

**Cardiospermum** L...................................................Sapindaceae

**Carex** L. .............................................................Cyperaceae

Careya see Planchonia, Barringtonia ............................Barringtoniaceae

Cargillia = Diospyros.................................................Ebenaceae

**Carica** L. ............................................................Caricaceae

**Cariniana** Casar.....................................................Barringtoniaceae

Carinta = Geophila ...................................................Rubiaceae

Carissa L. ...................................................... Apocynaceae
Carmona = Ehretia ................................................ Boraginaceae
Carpentaria Becc................................................... Arecaceae ................... no record
Carpha Banks & Sol. ex R. Br. ..................................... Cyperaceae
Carpodetus Forster & Forster f.................................... Grossulariaceae
Carpoxylon H. Wendl. & Drude ...................................... Arecaceae ................... no record
Carronia F. Muell. ............................................... Menispermaceae
Carruthersia Seemann ............................................. Apocynaceae
Carteretia = Cleisostoma.......................................... Orchidaceae
Carumbium see Homalanthus, Excoecaria........................... Euphorbiaceae
Caryodaphnopsis Airy Shaw ........................................ Lauraceae
Caryota L. ....................................................... Arecaceae
Casearia Jacq..................................................... Flacourtiaceae
Cassia L. ........................................................ Caesalpiniaceae
Cassidispermum = Burckella........................................ Sapotaceae
Cassine L. ....................................................... Celastraceae
Cassytha L. ...................................................... Lauraceae
Castanopsis (D. Don) Spach ....................................... Fagaceae
Castanospermum A. Cunn. ex Hook.................................. Fabaceae
Castilla Sessé ................................................... Moraceae
Castilloa = Castilla ............................................. Moraceae
Casuarina L. ..................................................... Casuarinaceae
Catanthera F. Muell. ............................................. Melastomataceae
Catharanthus G. Don f............................................. Apocynaceae
Cathormion (Benth.) Hassk. ....................................... Mimosaceae
Catimbium = Alpinia .............................................. Zingiberaceae
Caturus = Malaisia ............................................... Moraceae
Cayratia Juss. ................................................... Vitaceae
Cecarria Barlow .................................................. Loranthaceae
Cedrela P. Browne (Toona) ........................................ Meliaceae
Ceiba Miller...................................................... Bombacaceae
Celastrus L....................................................... Celastraceae
Celosia L. ....................................................... Amaranthaceae
Celtis L.......................................................... Ulmaceae
Cenchrus L........................................................ Poaceae
Centella L. ...................................................... Apiaceae
Centipeda Lour. .................................................. Asteraceae
Centotheca Desv................................................... Poaceae
Centranthera R. Br. .............................................. Scrophulariaceae
Centrolepis Labill................................................ Centrolepidaceae
Centrosema (DC.) Benth. .......................................... Fabaceae
Centrostigma Schltr. ............................................. Orchidaceae
Ceodes Forster & Forster f........................................ Nyctaginaceae

Cephaelis Sw. ...................................................... Rubiaceae
Cephalohibiscus Ulbr. = Thespesia .............................. Malvaceae
Cephalomanes C. Presl........................................... Hymenophyllaceae
Cephaloscirpus = Mapania......................................... Cyperaceae
Cerasiocarpum = Kedrostis ...................................... Cucurbitaceae
Cerastium L. ....................................................... Caryophyllaceae
Ceratanthus F. Muell. ex G. Taylor............................. Lamiaceae
Ceratochilus Blume .............................................. Orchidaceae
Ceratopetalum Sm. ................................................ Cunoniaceae
Ceratophyllum L. .................................................. Ceratophyllaceae
Ceratopteris Brongn. ............................................. Parkeriaceae
Ceratostylis Blume ............................................... Orchidaceae
Cerbera L. ......................................................... Apocynaceae
Ceriops Arn......................................................... Rhizophoraceae
Ceropegia L......................................................... Asclepiadaceae
Cerosora (Baker) Domin ......................................... Adiantaceae (Adiant.) ..... no record
Cestichis = Liparis ............................................... Orchidaceae
Cestrum L. ......................................................... Solanaceae
Ceuthostoma L. Johnson........................................... Casuarinaceae
Chaerefolium = Anthriscus ...................................... Apiaceae
Chaetochloa = Setaria ........................................... Poaceae
Chaetospora see Cladium, Rhynchospora, Schoenus ............ Cyperaceae
Chaetostachydium Airy Shaw .................................... Rubiaceae
Chaetostachys = Chaetostachydium............................... Rubiaceae
Chaetosus = Parsonsia............................................. Apocynaceae
Chailletia = Dichapetalum ....................................... Dichapetalaceae
Chalcas = Murraya................................................. Rutaceae
Chamaeanthus Schltr. ex J.J. Sm. .............................. Orchidaceae
Chamaeraphis see Pseudoraphis ................................. Poaceae
Chambeyronia Vieill. ............................................. Arecaceae ................... no record
Champereia Griffith .............................................. Opiliaceae
Chariessa = Citronella............................................ Icacinaceae
Chassalia Comm. ex Poiret ..................................... Rubiaceae
Cheilanthes Sw. ................................................... Adiantaceae (Adiant.)
Cheirodendron Nutt. ex Seemann ............................... Araliaceae
Cheiropleuria C. Presl. ......................................... Cheiropleuriaceae
Cheirostylis Blume ............................................... Orchidaceae
Chelonespermum = Burckella...................................... Sapotaceae
Chenopodium L...................................................... Chenopodiaceae
Chilocarpus Blume ................................................ Apocynaceae
Chilopogon = Appendicula ....................................... Orchidaceae
Chingia Holttum ................................................... Thelypteridaceae
Chionachne R. Br. ................................................ Poaceae
Chionanthus L. (Linociera)....................................... Oleaceae

**Chionochloa** Zotov ............................................. Poaceae
**Chisocheton** Blume ............................................. Meliaceae
**Chitonanthera** Schltr. ......................................... Orchidaceae
**Chitonochilus** Schltr. ......................................... Orchidaceae
**Chlaenandra** Miq. ............................................. Menispermaceae
Chloothamnus see Nastus ...................................... Poaceae
**Chloranthus** Sw. .............................................. Chloranthaceae
**Chloris** Sw. ................................................. Poaceae
Chlorocyperus = Cyperus ...................................... Cyperaceae
Chlorophora = Milicia .......................................... Moraceae
**Chlorophytum** Ker-Gawler .................................. Liliaceae ..................... no record
Chomelia = Tarenna ........................................... Rubiaceae
**Choriceras** Baillon (Dissilaria) ............................. Euphorbiaceae
**Chorizandra** R. Br. .......................................... Cyperaceae
**Christella** A. Léveillé ........................................ Thelypteridaceae
**Christensenia** Maxon ......................................... Marattiaceae
**Christia** Moench ............................................. Fabaceae
Christopteris Copel. (Christiopteris) .......................... Polypodiaceae (Micro.) ... no record
**Chrysanthemum** L. ........................................... Asteraceae
Chrysodium see Acrostichum ................................... Adiantaceae (Pter.)
**Chrysoglossum** Blume ........................................ Orchidaceae
**Chrysophyllum** L. ............................................ Sapotaceae
**Chrysopogon** Trin. ........................................... Poaceae
**Chydenanthus** Miers ......................................... Barringtoniaceae
**Cibotium** Kaulf. ............................................. Thyrsopteridaceae
Cinchona see Badusa ........................................... Rubiaceae
**Cinnamomum** Schaeffer ...................................... Lauraceae
Cirrhopetalum = Bulbophyllum ................................ Orchidaceae
**Cissampelos** L. .............................................. Menispermaceae
Cissodendron see Schefflera .................................... Araliaceae
**Cissus** L. ................................................... Vitaceae
**Citriobatus** A. Cunn. & Putterl. ............................. Pittosporaceae
**Citronella** D. Don ........................................... Icacinaceae
**Citrullus** Schrader .......................................... Cucurbitaceae
**Citrus** L. ................................................... Rutaceae
**Claderia** Hook. f. ........................................... Orchidaceae
**Cladium** P. Browne .......................................... Cyperaceae
**Cladomyza** Danser ........................................... Santalaceae
**Claoxylon** A. Juss. .......................................... Euphorbiaceae
Clarorivinia = Ptychopyxis ..................................... Euphorbiaceae
**Clausena** Burm. f. ........................................... Rutaceae
**Cleidion** Blume .............................................. Euphorbiaceae
**Cleisostoma** Blume (Sarcochilus) ............................ Orchidaceae

Cleistanthus Hook. f. ex Planchon ............................... Euphorbiaceae

Cleistocalyx = Syzygium ............................................ Myrtaceae

Cleistochloa C.E. Hubb. ........................................... Poaceae

Cleistopholis Pierre ex Engl. ...................................... Annonaceae

Clematis L. ......................................................... Ranunculaceae

Cleome L. ........................................................... Capparaceae

Clerodendrum L. ................................................... Verbenaceae

Clethra L. .......................................................... Clethraceae

Clidemia D. Don ................................................... Melastomataceae

Clinogyne see Donax, Marantochloa, Schumannianthus ........ Marantaceae

Clinostigma H. Wendl. ............................................. Arecaceae

Clitandropsis = Melodinus ......................................... Apocynaceae

Clitoria L. .......................................................... Fabaceae

Clymenia Swingle .................................................. Rutaceae

Coccinia Wight & Arn. .............................................. Cucurbitaceae

Coccoceras = Mallotus ............................................. Euphorbiaceae

Coccoglochidion = Glochidion ..................................... Euphorbiaceae

Coccoloba P. Browne (Muehlenbeckia) .......................... Polygonaceae

Cochlospermum Kunth .............................................. Bixaceae

Cocos L. ............................................................ Arecaceae

Codiaeum A. Juss. .................................................. Euphorbiaceae

Codonopsis Wallich ................................................ Campanulaceae

Codonosiphon Schltr. .............................................. Orchidaceae

Coelachne R. Br. ................................................... Poaceae

Coelodiscus = Mallotus ............................................. Euphorbiaceae

Coelogyne Lindley .................................................. Orchidaceae

Coelopyrena Valeton ................................................ Rubiaceae

Coelorachis Brongn. (Mnesithea) .................................. Poaceae

Coelospermum Blume (Caelospermum) .......................... Rubiaceae

Coffea L. ........................................................... Rubiaceae

Coix L. ............................................................. Poaceae

Coldenia L. ......................................................... Boraginaceae

Coleospadix = Drymophloeus ...................................... Arecaceae

Coleus = Solenostemon ............................................. Lamiaceae

Collabium Blume ................................................... Orchidaceae

Collyris = Dischidia ................................................ Asclepiadaceae

Colocasia Schott .................................................... Araceae

Colona Cav. ........................................................ Tiliaceae

Colubrina Rich. ex Brongn. ........................................ Rhamnaceae

Columbia = Colona ................................................. Tiliaceae

Colysis C. Presl. ................................................... Polypodiaceae (Micro.)

Combretopsis = Lophopyxis ........................................ Celastraceae

Combretum Loefl. .................................................. Combretaceae

Comesperma Labill. (Bredemeyera) ............................... Polygalaceae ............... no record

**Cominsia** Hemsley............................................... Marantaceae
**Commelina** L................................................ Commelinaceae
**Commersonia** Forster & Forster f................................ Sterculiaceae
**Conandrium** (Schumann) Mez ................................... Myrsinaceae
**Conchophyllum** Blume ...................................... Asclepiadaceae
**Coniogramme** Fée ................................... Adiantaceae (Adiant.)
**Connarus** L................................................. Connaraceae
Conocephalus = Poikilospermum (Cecropiaceae) ............... Urticaceae
Convolvulus Juss............................................ Convolvulaceae ............ no record
**Conyza** Less. ................................................ Asteraceae
Coombea P. Royen = Medicosma ................................ Rutaceae
**Coprosma** Forster & Forster f. ............................... Rubiaceae
Coptophyllum Korth........................................... Rubiaceae.................... no record
**Coptosapelta** Korth.......................................... Rubiaceae
**Corchorus** L. ............................................... Tiliaceae
**Cordia** L.................................................. Boraginaceae
**Cordyline** Comm. ex R. Br. ................................... Agavaceae
Coreopsis L.................................................. Asteraceae ................. no record
**Coriandrum** L. .............................................. Apiaceae
**Coriaria** L................................................ Coriariaceae
Coridochloa = Alloteropsis.................................... Poaceae
Cornopteris Nakai............................................ Aspleniaceae (Athyr.)..... no record
Cornutia see Premna.......................................... Verbenaceae
**Coronanthera** Vieill. ex C.B. Clarke ............................ Gesneriaceae
**Corsia** Becc. ............................................... Corsiaceae
**Corybas** Salisb. ............................................. Orchidaceae
Corymbis = Corymborkis....................................... Orchidaceae
**Corymborkis** Thouars ...................................... Orchidaceae
**Corynocarpus** Forster & Forster f. ........................... Corynocarpaceae
**Corypha** L. ............................................... Arecaceae
**Coryphopteris** Holttum..................................... Thelypteridaceae
Corysanthes = Corybas......................................... Orchidaceae
**Cosmos** Cav................................................. Asteraceae
**Costularia** C.B. Clarke ex Dyer ................................ Cyperaceae
**Costus** L. ................................................ Zingiberaceae
**Cotula** L. ................................................ Asteraceae
**Cotylanthera** Blume ........................................ Gentianaceae
Couthovia = Neuburgia ........................................ Loganiaceae
Craspedodictyum = Syngramma................................ Adiantaceae (Adiant.)
**Crassocephalum** Moench ..................................... Asteraceae
Crateriphytum = Neuburgia ..................................... Loganiaceae
**Crateva** L. (Crataeva)........................................ Capparaceae
Cratoxylum Blume ............................................ Clusiaceae ................. no record

Cremnobates = Schizomeria ........................................ Cunoniaceae

**Creochiton** Blume ..................................................... Melastomataceae

**Crepidomanes** (C. Presl) C. Presl (Trichomanes).............. Hymenophyllaceae

Crepidopteris see Crepidomanes .................................... Hymenophyllaceae

Crepis L........................................................................ Asteraceae .................. no record

**Crescentia** L. .............................................................. Bignoniaceae

**Crinum** L.................................................................... Amaryllidaceae

**Crossandra** Salisb. ..................................................... Acanthaceae

Crossonephelis = Glenniea ............................................ Sapindaceae

**Crossostylis** Forster & Forster f. ................................. Rhizophoraceae

**Crotalaria** L. .............................................................. Fabaceae

**Croton** L.................................................................... Euphorbiaceae

**Crucicaryum** Brand .................................................... Boraginaceae

**Crudia** Schreber ......................................................... Caesalpiniaceae

**Crypsinus** C. Presl. ..................................................... Polypodiaceae (Micro.)

**Crypteronia** Blume...................................................... Crypteroniaceae

**Cryptocarya** R. Br. ..................................................... Lauraceae

**Cryptocoryne** Fischer ex Wydler ................................. Araceae

Cryptogramma R. Br. ..................................................... Adiantaceae (Adiant.)..... no record

**Cryptostegia** R. Br. ..................................................... Asclepiadaceae

**Cryptostylis** R. Br........................................................ Orchidaceae

**Ctenitis** (C. Chr.) Tard. & C. Chr. ................................ Aspleniaceae (Tect.)

**Ctenolophon** Oliver ................................................... Linaceae

**Ctenopteris** Blume ex Kunze ....................................... Grammitidaceae

Cubilia Blume ............................................................... Sapindaceae ................ no record

**Cucumis** L. ................................................................ Cucurbitaceae

**Cucurbita** L................................................................ Cucurbitaceae

**Cudrania** Trécul (Maclura) .......................................... Moraceae

**Culcita** C. Presl. .......................................................... Thyrsopteridaceae

Cullen see Psoralea ...................................................... Fabaceae

Cumingia = Camptostemon ........................................... Bombacaceae

Cunonia see Caldcluvia .................................................. Cunoniaceae

Cupania see Arytera, Nephelium ................................... Sapindaceae

**Cupaniopsis** Radlk. ..................................................... Sapindaceae

**Curculigo** Gaertner ..................................................... Hypoxidaceae

**Curcuma** Roxb. ........................................................... Zingiberaceae

Currania = Gymnocarpium ............................................ Aspleniaceae (Athyr.)

**Cuscuta** L. ................................................................. Convolvulaceae

**Cyanotis** D. Don ......................................................... Commelinaceae

**Cyathea** Sm. ............................................................... Cyatheaceae

**Cyathocalyx** Champ. ex Hook f. & Thomson.................... Annonaceae

**Cyathostemma** Griff. ................................................... Annonaceae

**Cyathula** Blume.......................................................... Amaranthaceae

**Cycas** L. .................................................................... Cycadaceae

Cyclandra = Ternstroemia ............................................. Theaceae

**Cyclandrophora** Hassk. (Atuna)................................... Chrysobalanaceae

Cyclobalanopsis see Lithocarpus .................................... Fagaceae

Cyclocampe see Lophoschoenus, Schoenus ...................... Cyperaceae

**Cyclopeltis** J. Sm. .................................................... Aspleniaceae (Tect.)

Cyclophorus = Pyrrosia................................................ Polypodiaceae (Platy.)

**Cyclosorus** Link ...................................................... Thelypteridaceae

Cylindrokelupha = Archidendron .............................. Mimosaceae

**Cymaria** Benth. ....................................................... Lamiaceae

**Cymbidium** Sw. ....................................................... Orchidaceae

**Cymbopogon** Sprengel................................................ Poaceae

**Cymodocea** König ..................................................... Cymodoceaceae

**Cynanchum** L. .......................................................... Asclepiadaceae

Cynoctonum = Mitreola................................................ Loganiaceae

**Cynodon** Rich. ......................................................... Poaceae

**Cynoglossum** L......................................................... Boraginaceae

**Cynometra** L. .......................................................... Caesalpiniaceae

**Cynorkis** Thouars...................................................... Orchidaceae

Cynosurus see Dactylocnemium, Eleusine ...................... Poaceae

**Cyperus** L. ............................................................. Cyperaceae

**Cyphochilus** Schltr. ................................................... Orchidaceae

**Cypholophus** Wedd. ................................................... Urticaceae

**Cyphomandra** C. Martius ex Sendtner........................... Solanaceae

Cyphophoenix H. Wendl. ex Hook. f. ........................... Arecaceae ................... no record

Cypripedium see Paphiopedilum ................................... Orchidaceae

**Cyrtandra** Forster & Forster f. ................................... Gesneriaceae

Cyrtandropsis = Tetraphyllum .................................... Gesneriaceae

**Cyrtococcum** Stapf (Panicum)...................................... Poaceae

Cyrtopera = Eulophia ................................................ Orchidaceae

**Cyrtorchis** Schltr. ...................................................... Orchidaceae

**Cyrtosperma** Griffith ................................................. Araceae

**Cyrtostachys** Blume ................................................... Arecaceae

**Cystodium** J. Sm....................................................... Cyatheaceae

**Cystopteris** Bernh. .................................................... Aspleniaceae (Athyr.)

Cystopus = Pristiglottis ............................................. Orchidaceae

**Cystorchis** Blume ...................................................... Orchidaceae

Cytisus see Pongamia ................................................. Fabaceae

**Dacrycarpus** (Endl.) Laubenf. (Podocarpus) ..................... Podocarpaceae

**Dacrydium** Lambert.................................................... Podocarpaceae

**Dacryodes** Vahl ........................................................ Burseraceae

**Dactyliophora** Tieghem (Dactylophora)........................... Loranthaceae

**Dactylocladus** Oliver .................................................. Crypteroniaceae

**Dactyloctenium** Willd.................................................. Poaceae

**Dactylorhynchus** Schltr. ............................................. Orchidaceae
**Daemonorops** Blume................................................. Arecaceae
**Dahlia** Cav. ........................................................ Asteraceae
Dais see Phaleria ..................................................... Thymelaeaceae
**Dalbergia** L. f. ..................................................... Fabaceae
**Dallachya** F. Muell. (Rhamnella)................................. Rhamnaceae
Dammara = Agathis.................................................. Araucariaceae
Dammaropsis = Ficus ................................................ Moraceae
Dammera = Licuala ................................................... Arecaceae
Dansera = Dialium ................................................... Caesalpiniaceae
**Danthonia** DC. ...................................................... Poaceae
Daphnandra see Dryadodaphne ..................................... Monimiaceae
**Daphniphyllum** Blume............................................... Daphniphyllaceae
Darea = Asplenium ................................................... Aspleniaceae (Asplen.)
Dasycoleum = Chisocheton........................................... Meliaceae
**Datura** L............................................................ Solanaceae
**Daucus** L. .......................................................... Apiaceae
**Davallia** Sm. ....................................................... Davalliaceae (Davall.)
**Davallodes** (Copel.) Copel. ....................................... Davalliaceae (Davall.)
**Debregeasia** Gaudich. ............................................. Urticaceae
**Decaisnina** Tieghem ................................................ Loranthaceae
**Decaspermum** Forster & Forster f. ............................... Myrtaceae
**Decatoca** F. Muell. ................................................ Epacridaceae
**Decussocarpus** Laubenf. (Nageia) ............................... Podocarpaceae
Dedea = Quintinia.................................................... Saxifragaceae
**Deeringia** R. Br. ................................................... Amaranthaceae
**Dehaasia** Blume .................................................... Lauraceae
**Delarbrea** Vieill.................................................... Araliaceae
**Delonix** Raf........................................................ Caesalpiniaceae
**Delphyodon** K. Schum. ............................................. Apocynaceae
**Dendrobium** Sw...................................................... Orchidaceae
Dendrocalamus see Bambusa ........................................ Poaceae
**Dendrochilum** Blume ............................................... Orchidaceae
**Dendrocnide** Miq. .................................................. Urticaceae
Dendrocolla = Sarcochilus .......................................... Orchidaceae
**Dendroconche** Copel. .............................................. Polypodiaceae (Micro.)
Dendroglossa = Colysis .............................................. Polypodiaceae (Micro.)
Dendrolobium (Wight & Arn.) Benth. see Desmodium ........ Fabaceae
**Dendromyza** Danser ................................................ Santalaceae
**Dendrophtoe** C. Martius ........................................... Loranthaceae
**Dendrotrophe** Miq.................................................. Santalaceae
**Dennstaedtia** Bernh. .............................................. Dennstaedtiaceae (Denn.)
**Dentella** Forster & Forster f. ..................................... Rubiaceae
**Deparia** Hook. & Grev............................................. Aspleniaceae (Athyr.)..... no record

**Deplanchea** Vieill. ...................................................... Bignoniaceae

**Derris** Lour. .............................................................. Fabaceae

**Deschampsia** Pal. ....................................................... Poaceae

**Desmanthus** Willd. (Neptunia) ...................................... Mimosaceae

**Desmodium** Desv. ........................................................ Fabaceae

**Desmos** Lour. ............................................................. Annonaceae

Desmotrichum = Flickingeria ......................................... Orchidaceae

**Detzneria** Schltr. ex Diels ............................................ Scrophulariaceae

Deyeuxia = Calamagrostis ............................................. Poaceae

**Diacalpe** Blume .......................................................... Aspleniaceae (Dryopt.)

**Dialium** L. ............................................................... Caesalpiniaceae ........... no record

Diandriella = Homalomena .............................................. Araceae

**Dianella** Lam. ............................................................ Liliaceae

**Diblemma** J. Sm. ........................................................ Polypodiaceae (Micro.)... no record

Dicerma see Desmodium .................................................. Fabaceae

**Dicerospermum** Bakh. f. ............................................... Melastomataceae

**Dichanthium** Willemet ................................................. Poaceae

**Dichapetalum** Thouars .................................................. Dichapetalaceae

**Dichelachne** Endl. ....................................................... Poaceae

**Dichondra** Forster & Forster f. ..................................... Convolvulaceae

Dichopogon = Arthropodium ........................................... Liliaceae

Dichopus = Dendrobium .................................................. Orchidaceae

**Dichroa** Lour. ............................................................ Hydrangeaceae

**Dichrocephala** L'Hérit. ex DC. ....................................... Asteraceae

**Dichrotrichum** Reinw. ex Vriese (Aeschynanthus) .............. Gesneriaceae

**Dicksonia** L'Hérit. ...................................................... Cyatheaceae

**Dicliptera** Juss. ......................................................... Acanthaceae

**Dicranopteris** Bernh. ................................................... Gleicheniaceae

**Dictymia** J. Sm. ......................................................... Polypodiaceae (Poly.)..... no record

Dictyocline T. Moore ..................................................... Thelypteridaceae ........... no record

**Dictyoneura** Blume ...................................................... Sapindaceae

Dictyopteris = Pleocnemia .............................................. Aspleniaceae (Tect.)

Dicymanthes = Amyema ................................................. Loranthaceae

Didiscus = Trachymene .................................................. Apiaceae

**Didissandra** C.B. Clarke .............................................. Gesneriaceae

Didymocarpus see Boea .................................................. Gesneriaceae

Didymocheton = Dysoxylum ............................................. Meliaceae

**Didymochlaena** Desv. ................................................... Aspleniaceae (Tect.)

**Didymoplexis** Griffith .................................................. Orchidaceae

Didymosperma = Arenga ................................................ Arecaceae

**Dieffenbachia** Schott .................................................. Araceae

Dienia = Malaxis ......................................................... Orchidaceae

Digastrium see Ischaemum .............................................. Poaceae

Digitaria Haller ...................................................... Poaceae

Diglyphosa Blume..................................................... Orchidaceae................. no record

Dillenia L.............................................................. Dilleniaceae

Dimeria R. Br. ....................................................... Poaceae

Dimocarpus Lour. (Litchi) ........................................ Sapindaceae

Dimorphanthera (Drude) F. Muell. ex J.J. Sm. ................. Ericaceae

Dimorphocalyx Thwaites............................................ Euphorbiaceae

Dinochloa Buese ..................................................... Poaceae

Dioclea Kunth......................................................... Fabaceae

Diodia L. .............................................................. Rubiaceae

Dioscorea L. .......................................................... Dioscoreaceae

Diospyros L. .......................................................... Ebenaceae

Diplacrum = Scleria.................................................. Cyperaceae

Diplanthera = Deplanchea .......................................... Bignoniaceae

Diplaziopsis C. Chr. (Diplazium)................................... Aspleniaceae (Athyr.)

Diplazium Sw. ........................................................ Aspleniaceae (Athyr.)

Diplectria (Blume) Reichb. ......................................... Melastomataceae

Diplocaulobium (Reichb. f.) Kraenzlin (Dendrobium) ......... Orchidaceae

Diploclisia Miers..................................................... Menispermaceae

Diplocyclos (Endl.) Post & Kuntze .............................. Cucurbitaceae

Diploglottis Hook. f. ................................................ Sapindaceae

Diplopterygium (Diels) Nakai ..................................... Gleicheniaceae

Diplora Baker (Asplenium) ........................................ Aspleniaceae (Asplen.)

Diplospora DC. (Tricalysia)........................................ Rubiaceae

Diplycosia Blume ..................................................... Ericaceae

Dipodium R. Br. ..................................................... Orchidaceae

Dipteracanthus = Ruellia............................................ Acanthaceae

Dipteris Reinw. ...................................................... Dipteridaceae

Discalyxia = Alyxia.................................................. Apocynaceae

Dischidia R. Br. ...................................................... Asclepiadaceae

Discocalyx (A. DC.) Mez ........................................... Myrsinaceae

Discogyne = Ixonanthes.............................................. Linaceae

Disiphon = Vaccinium ............................................... Ericaceae

Disperis Sw. .......................................................... Orchidaceae

Dissiliaria see Choriceras ........................................... Euphorbiaceae

Dissochaeta Blume.................................................... Melastomataceae

Distemon see Neodistemon........................................... Urticaceae

Distreptus = Elephantopus .......................................... Asteraceae

Distrianthes Danser .................................................. Loranthaceae

Distyliopsis see Sycopsis ............................................ Hamamelidaceae

Dodonaea Miller ...................................................... Sapindaceae

Dolianthus = Amaracarpus .......................................... Rubiaceae

Dolichandrone (Fenzl) Seemann.................................... Bignoniaceae

Dolicholobium A. Gray .............................................. Rubiaceae

Dolichos see Canavalia, Lablab, Vigna ............................ Fabaceae

**Donax** Lour. ................................................... Marantaceae

**Doodia** R. Br. .................................................. Blechnaceae

**Doritis** Lindley ................................................. Orchidaceae

**Doryopteris** J. Sm. ............................................. Adiantaceae (Adiant.)

**Dossinia** Morren ................................................ Orchidaceae

**Dovyalis** E. Meyer ex Arn. ..................................... Flacourtiaceae

**Dracaena** Vand. ex L. ........................................... Dracaenaceae

**Dracontomelon** Blume ........................................... Anacardiaceae

**Drakaea** Lindley ................................................ Orchidaceae

**Drapetes** Lam. .................................................. Thymelaeaceae

Drepananthus = Cyathocalyx ...................................... Annonaceae

**Drimys** Forster & Forster f. .................................... Winteraceae

Drimyspermum = Phaleria ........................................ Thymelaeaceae

**Drosera** L. ..................................................... Droseraceae

**Dryadodaphne** S. Moore ......................................... Monimiaceae

**Dryadorchis** Schltr. ............................................ Orchidaceae

**Drymaria** Willd. ex Schultes ................................... Caryophyllaceae

Drymoglossum = Pyrrosia ......................................... Polypodiaceae (Platy.)

**Drymophloeus** Zipp. ............................................ Arecaceae

**Drynaria** (Bory) J. Sm. ......................................... Polypodiaceae (Poly.)

**Drynariopsis** (Copel.) Ching ................................... Polypodiaceae (Poly.)

Dryoathyrium = Deparia .......................................... Aspleniaceae (Athyr.)

**Dryopolystichum** Copel. ........................................ Aspleniaceae (Tect.)

**Dryopteris** Adans. .............................................. Aspleniaceae (Dryopt.)

Dryostachyum = Aglaomorpha ..................................... Polypodiaceae (Dryn.)

**Drypetes** Vahl ................................................. Euphorbiaceae

**Duabanga** Buch.-Ham. ........................................... Sonneratiaceae

**Dubouzetia** Pancher ex Brongn. & Gris ......................... Elaeocarpaceae

Duckera = Rhus .................................................. Anacardiaceae

**Dumasia** DC. ................................................... Fabaceae

**Dunbaria** Wight & Arn. ......................................... Fabaceae

**Durandea** Planchon (Hugonia) .................................. Linaceae

**Duranta** L. .................................................... Verbenaceae

**Durio** Adans. .................................................. Bombacaceae

Duval-Jouvea = Cyperus ........................................... Cyperaceae

**Dysophylla** Blume (Pogostemon) ................................ Lamiaceae

**Dysoxylum** Blume ............................................... Meliaceae

**Earina** Lindley ................................................ Orchidaceae

Ecdysanthera Hook. & Arn. ....................................... Apocynaceae ............... no record

**Echinocarpus** Blume (Sloanea) ................................. Elaeocarpaceae

**Echinochloa** Pal. .............................................. Poaceae

Echinocitrus Tanaka = Triphasia ................................. Rutaceae

Echinolaena Desv. (Pseudechinolaena) ............................ Poaceae
Echinopogon Pal. ..................................................... Poaceae
Echinospermum = Lappula ......................................... Boraginaceae
Eclipta L. ............................................................... Asteraceae
Ectrosia R. Br. ........................................................ Poaceae
Ectrosiopsis (Ohwi) Jansen .......................................... Poaceae ..................... no record
Egenolfia = Bolbitis ................................................. Aspleniaceae (Lom.)
Ehretia P. Browne .................................................... Boraginaceae
Ehrharta Thunb. ...................................................... Poaceae
Eichhornia Kunth ..................................................... Pontederiaceae
Elaeagnus L. ........................................................... Elaeagnaceae
Elaeis Jacq. ............................................................ Arecaceae
Elaeocarpus L. ........................................................ Elaeocarpaceae
Elaeodendron = Cassine .............................................. Celastraceae
Elaphoglossum Schott ex J. Sm. ................................... Aspleniaceae (Elaph.)
Elatine L. ............................................................... Elatinaceae
Elatostema Forster & Forster f. ..................................... Urticaceae
Elattostachys Radlk. ................................................. Sapindaceae
Eleocharis R. Br. ..................................................... Cyperaceae
Elephantopus L. ...................................................... Asteraceae
Elettaria Maton ....................................................... Zingiberaceae
Elettariopsis Baker ................................................... Zingiberaceae
Eleusine Gaertner ..................................................... Poaceae
Eleutheranthera Poit. ex Bosc. ..................................... Asteraceae
Eleutherostylis Burret ................................................ Tiliaceae
Elionurus Humb. & Bonpl. ex Willd. ............................. Poaceae
Ellisiophyllum Maxim. ............................................... Scrophulariaceae
Elmerrillia Dandy .................................................... Magnoliaceae
Elymus L. .............................................................. Poaceae
Elyonurus = Elionurus ............................................... Poaceae
Elytranthe see Decaisnina, Macrosolen ........................... Loranthaceae
Embelia Burm. f. ..................................................... Myrsinaceae
Embothrium see Oreocallis .......................................... Proteaceae
Emilia Cass. ........................................................... Asteraceae
Emmenosperma F. Muell. ........................................... Rhamnaceae
Endiandra R. Br. ..................................................... Lauraceae
Endocomia Wilde (Horsfieldia) ..................................... Myristicaceae
Endospermum Benth. ................................................ Euphorbiaceae
Engelhardia Leschen. ex Blume (Engelhardtia) ................. Juglandaceae
Enhalus Rich. ......................................................... Hydrocharitaceae
Enkleia Griffith ....................................................... Thymelaeaceae
Enneapogon Desv. ex Pal. .......................................... Poaceae
Ensete Horan. ......................................................... Musaceae
Entada Adans. ......................................................... Mimosaceae

**Enterolobium** C. Martius ............................................ Mimosaceae
**Enteropogon** Nees ..................................................... Poaceae
Entolasia see Panicum ................................................ Poaceae
Epacris Cav.............................................................. Epacridaceae ............... no record
**Epaltes** Cass. ......................................................... Asteraceae
Ephemerantha = Flickingeria......................................... Orchidaceae
Ephippium = Bulbophyllum ............................................ Orchidaceae
**Epiblastus** Schltr. .................................................... Orchidaceae
Epicharis = Dysoxylum ............................................... Meliaceae
**Epidendrum** L. ........................................................ Orchidaceae
**Epigeneium** Gagnepain (Dendrobium) ............................ Orchidaceae
**Epilobium** L. .......................................................... Onagraceae
Epimeredi Adans. ...................................................... Lamiaceae .................. no record
**Epipactis** Zinn (Goodyera) ......................................... Orchidaceae
**Epipogium** J. Gmelin ex Borkh..................................... Orchidaceae
Epipremnopsis = Amydrium ........................................... Araceae
**Epipremnum** Schott................................................... Araceae
Epirixanthes = Salomonia.............................................. Polygalaceae
**Epithema** Blume ....................................................... Gesneriaceae
**Equisetum** L............................................................ Equisetaceae
**Eragrostis** Wolf ...................................................... Poaceae
**Eranthemum** L. ....................................................... Acanthaceae
**Erechtites** Raf. ....................................................... Asteraceae
**Eremochloa** Buese..................................................... Poaceae
**Eria** Lindley........................................................... Orchidaceae
**Eriachne** R. Br. ....................................................... Poaceae
**Eriandra** P. Royen & Steenis ...................................... Polygalaceae
Erianthus = Saccharum................................................. Poaceae
**Erigeron** L. ........................................................... Asteraceae
Erinus see Poarium .................................................... Scrophulariaceae
**Eriocaulon** L. ......................................................... Eriocaulaceae
**Eriochloa** Kunth ...................................................... Poaceae
Eriodendron = Ceiba.................................................... Bombacaceae
Erioglossum = Lepisanthes ........................................... Sapindaceae
Eriolopha = Alpinia..................................................... Zingiberaceae
**Eriosema** (DC.) G. Don f. .......................................... Fabaceae
Erpetina = Medinilla ................................................... Melastomataceae
Ervatamia = Tabernaemontana......................................... Apocynaceae
**Erycibe** Roxb. ........................................................ Convolvulaceae
**Erythrina** L............................................................ Fabaceae
**Erythrodes** Blume ..................................................... Orchidaceae
**Erythropalum** Blume ................................................. Olacaceae
**Erythrospermum** Lam.................................................. Flacourtiaceae

Erythroxylum P. Browne ............................................ Erythroxylaceae
Eschweileria see Osmoxylon (Araliaceae) ......................... Barringtoniaceae
Ethulia L. f. ....................................................... Asteraceae
Etlingera = Amomum ............................................... Zingiberaceae
Eucalyptopsis C. White ............................................ Myrtaceae
Eucalyptus L'Hérit. ............................................... Myrtaceae
Eucharis Planchon & Linden ....................................... Amaryllidaceae
Eucosia Blume .................................................... Orchidaceae
Eugenia L. ....................................................... Myrtaceae
Eulalia Kunth .................................................... Poaceae
Eulophia R. Br. ................................................... Orchidaceae
Euodia J.R. & G. Forst. (Melicope) ............................... Rutaceae
Euonymus L. ...................................................... Celastraceae
Euphorbia L. ..................................................... Euphorbiaceae
Euphoria = Dimocarpus............................................ Sapindaceae
Euphorianthus = Diploglottis ..................................... Sapindaceae
Euphrasia L....................................................... Scrophulariaceae
Euplassa see Gevuina ............................................. Proteaceae
Eupomatia R. Br................................................... Eupomatiaceae
Euroschinus Hook. f. ............................................. Anacardiaceae
Eurya Thunb....................................................... Theaceae
Eurycentrum Schltr. .............................................. Orchidaceae
Eurycles = Proiphys .............................................. Amaryllidaceae
Eustrephus R. Br. ................................................ Smilacaceae
Euthamnus see Aeschynanthus....................................... Gesneriaceae
Euxolus = Amaranthus ............................................. Amaranthaceae
Everettia see Astronidium ........................................ Melastomataceae
Evia = Spondias .................................................. Anacardiaceae
Evodia = Euodia .................................................. Rutaceae
Evodiella Van der Linden ......................................... Rutaceae
Evolvulus L....................................................... Convolvulaceae
Exacum L. ........................................................ Gentianaceae
Excavatia = Ochrosia ............................................. Apocynaceae
Excoecaria L. .................................................... Euphorbiaceae
Exocarpos Labill.................................................. Santalaceae
Exocarya Benth. .................................................. Cyperaceae
Fagara = Zanthoxylum ............................................. Rutaceae
Fagraea Thunb. ................................................... Loganiaceae
Fagraeopsis = Mastixiodendron .................................... Rubiaceae
Fahrenheitia Reichb. f. & Zoll. ex Muell. Arg. (Ostodes)..... Euphorbiaceae
Faikea Philipson (Faika).......................................... Monimiaceae
Falcaria Bernh. (Oenanthe)........................................ Apiaceae
Falcatifolium Laubenf. (Podocarpus) .............................. Podocarpaceae
Faradaya F. Muell. ............................................... Verbenaceae

**Fatoua** Gaudich. ...................................... Moraceae

Felicia Cass. ...................................... Asteraceae .................. no record

Fenzlia = Myrtella ...................................... Myrtaceae

**Festuca** L. ...................................... Poaceae

**Ficus** L. ...................................... Moraceae

**Fimbristylis** Vahl ...................................... Cyperaceae

**Finlaysonia** Wallich ...................................... Asclepiadaceae

**Finschia** Warb. ...................................... Proteaceae

**Firmiana** Marsili...................................... Sterculiaceae

**Fissistigma** Griffith ...................................... Annonaceae

**Fittingia** Mez ...................................... Myrsinaceae

**Flacourtia** Comm. ex L'Hérit. ...................................... Flacourtiaceae

**Flagellaria** L...................................... Flagellariaceae

**Flemingia** Roxb. ex Aiton & Aiton f. ...................................... Fabaceae

Fleurya = Laportea ...................................... Urticaceae

**Flickingeria** A. Hawkes ...................................... Orchidaceae

**Flindersia** R. Br...................................... Rutaceae

**Floscopa** Lour. ...................................... Commelinaceae

Flueggea see Securinega ...................................... Euphorbiaceae

Flueggeopsis = Phyllanthus ...................................... Euphorbiaceae

Folium see Heliconia ...................................... Heliconiaceae

**Fontainea** Heckel ...................................... Euphorbiaceae

Forrestia = Amischotolype ...................................... Commelinaceae

Fourcroya = Furcraea ...................................... Agavaceae

**Fragaria** L. ...................................... Rosaceae

**Freycinetia** Gaudich. ...................................... Pandanaceae

**Friesodielsia** Steenis...................................... Annonaceae

**Fuirena** Rottb...................................... Cyperaceae

**Furcraea** Vent...................................... Agavaceae

Gaertnera Lam. ...................................... Rubiaceae.................... no record

**Gahnia** Forster & Forster f. ...................................... Cyperaceae

**Gaimardia** Gaudich. ...................................... Centrolepidaceae

Gajanus = Inocarpus ...................................... Fabaceae

**Galactia** P. Browne ...................................... Fabaceae

**Galbulimima** Bailey ...................................... Himantandraceae

**Galearia** Zoll. & Moritzi...................................... Pandaceae

**Galeola** Lour. ...................................... Orchidaceae

**Galinsoga** Ruíz & Pavón ...................................... Asteraceae

**Galium** L. ...................................... Rubiaceae

**Ganophyllum** Blume...................................... Sapindaceae

Ganua = Madhuca ...................................... Sapotaceae

**Garcinia** L...................................... Clusiaceae

**Gardenia** Ellis...................................... Rubiaceae

# I. List of the Genera in New Guinea and the Solomon Islands

Garnotia Brongn. ..................................................... Poaceae

Garuga Roxb. ..................................................... Burseraceae

Gastonia Comm. ex Lam............................................ Araliaceae

Gastrodia R. Br. ..................................................... Orchidaceae

Gastroglottis = Liparis ............................................. Orchidaceae

Gastrolepis Tieghem ..................................... Icacinaceae ................. no record

Gaultheria L. ..................................................... Ericaceae

Geanthus = Amomum ............................................... Zingiberaceae

Geijera Schott ..................................................... Rutaceae

Geissanthera = Microtatorchis ..................................... Orchidaceae

Geissois Labill. ..................................................... Cunoniaceae

Geitonoplesium Cunn. ex R. Br. ................................. Smilacaceae

Gelibia = Polyscias..................................................... Araliaceae

Gelonium = Suregada..................................................... Euphorbiaceae

Gendarussa Nees (Justicia) ...................................... Acanthaceae

Geniostoma Forster & Forster f. ................................. Loganiaceae

Gentiana L. ..................................................... Gentianaceae

Geocardia = Geophila ............................................. Rubiaceae

Geodorum Jackson ..................................................... Orchidaceae

Geophila D. Don ..................................................... Rubiaceae

Geranium L. ..................................................... Geraniaceae

Germainia Bal. & Poitr. ............................................. Poaceae

Gertrudia = Ryparosa ............................................. Flacourtiaceae

Gestroa = Erythrospermum............................................ Flacourtiaceae

Geunsia = Callicarpa ............................................. Verbenaceae

Gevuina Molina ..................................................... Proteaceae

Gibbsia Rendle ..................................................... Urticaceae

Gigantochloa Kurz ex Munro ...................................... Poaceae

Gigasiphon Drake (Bauhinia)........................................ Caesalpiniaceae

Gillbeea F. Muell. ..................................................... Cunoniaceae

Ginalloa Korth. ..................................................... Viscaceae

Girardinia Gaudich..................................................... Urticaceae

Gironniera Gaudich. ..................................................... Ulmaceae

Giulianettia = Glossorhyncha ...................................... Orchidaceae

Gjellerupia Lauterb. ............................................. Opiliaceae

Glabraria = Brownlowia ............................................. Tiliaceae

Gladiolus L. ..................................................... Iridaceae

Gleditsia L..................................................... Caesalpiniaceae ............ no record

Gleichenia Sm. ..................................................... Gleicheniaceae

Glenniea Hook. f..................................................... Sapindaceae

Glinus L. ..................................................... Molluginaceae

Gliricidia Kunth..................................................... Fabaceae

Globba L. ..................................................... Zingiberaceae

Glochidion Forster & Forster f. ...................................... Euphorbiaceae

**Glomera** Blume ........................................... Orchidaceae
**Gloriosa** L. ........................................... Liliaceae
**Glossocarya** Wallich ex Griffith ..................................... Verbenaceae
Glossogyne see Bidens ............................................ Asteraceae
**Glossorhyncha** Ridley (Glomera) ................................. Orchidaceae
**Gluta** L. ........................................... Anacardiaceae
**Glycine** Willd. ........................................... Fabaceae
**Glycosmis** Corr. Serr. ........................................... Rutaceae
**Gmelina** L. ........................................... Verbenaceae
**Gnaphalium** L. ........................................... Asteraceae
**Gnetum** L. ........................................... Gnetaceae
Gomozia = Nertera........................................... Rubiaceae
**Gomphandra** Wallich ex Lindley ................................ Icacinaceae
Gomphocarpus = Asclepias........................................... Asclepiadaceae
**Gompholobium** Sm........................................... Fabaceae
**Gomphrena** L. ........................................... Amaranthaceae
**Gongronema** (Endl.) Decne. ........................................... Asclepiadaceae
Goniocarpus = Haloragis ........................................... Haloragidaceae
Goniocheton = Dysoxylum ........................................... Meliaceae
**Goniophlebium** C. Presl. ........................................... Polypodiaceae (Poly.)
Goniopteris see Cyclosorus, Dryopteris ........................... Thelypteridaceae
**Goniothalamus** (Blume) Hook. f. & Thomson .................. Annonaceae
**Gonocaryum** Miq. ........................................... Icacinaceae
**Gonocormus** Bosch (Trichomanes) ............................... Hymenophyllaceae
Gonostegia = Pouzolzia ........................................... Urticaceae
**Gonystylus** Teijsm. & Binnend. ................................ Thymelaeaceae
**Goodenia** Sm. ........................................... Goodeniaceae
**Goodyera** R. Br. ........................................... Orchidaceae
**Gordonia** Ellis........................................... Theaceae
Gossampinus see Bombax ........................................... Bombacaceae
**Gossypium** L. ........................................... Malvaceae
**Gouania** Jacq. ........................................... Rhamnaceae
**Gouldia** A. Gray (Psychotria) ................................ Rubiaceae
**Grammatophyllum** Blume........................................... Orchidaceae
**Grammatopteridium** Alderw. ................................... Polypodiaceae (Micro.)
**Grammitis** SW. ........................................... Grammitidaceae
**Graptophyllum** Nees ........................................... Acanthaceae
Gratiola see Lindernia........................................... Scrophulariaceae
**Grenacheria** Mez ........................................... Myrsinaceae
**Grevillea** R. Br. ex J. Knight........................................... Proteaceae
**Grewia** L. ........................................... Tiliaceae
Grisebachia see Laccospadix ........................................... Arecaceae
**Gronophyllum** R. Scheffer ........................................... Arecaceae

Grumilea = Psychotria..................................................Rubiaceae

**Guettarda** L.................................................................Rubiaceae

**Guettardella** Champ. ex Benth. (Anthirea) ......................Rubiaceae

Guilandina = Caesalpinia ............................................Caesalpiniaceae

Guillainea Vieill. = Alpinia .......................................Zingiberaceae

**Guioa** Cav. ...............................................................Sapindaceae

**Gulubia** Becc. ..........................................................Arecaceae

Gulubiopsis Becc. = Gulubia.......................................Arecaceae

**Gunnera** L. .................................................................Gunneraceae

**Gustavia** L. ................................................................Barringtoniaceae

**Gymnacranthera** Warb. ...............................................Myristicaceae

**Gymnanthera** R. Br. ....................................................Asclepiadaceae

**Gymnema** R. Br. ..........................................................Asclepiadaceae

**Gymnocarpium** Newman ..............................................Aspleniaceae (Athyr.)

Gymnogramma = Gymnopteris ....................................Adiantaceae (Adiant.)

**Gymnopetalum** Arn.......................................................Cucurbitaceae

**Gymnophragma** Lindau................................................Acanthaceae

Gymnopogon Pal. .......................................................Poaceae .....................no record

**Gymnopteris** Bernh. (Hemionotis) ...............................Adiantaceae (Adiant.)

**Gymnosiphon** Blume....................................................Burmanniaceae

Gymnosphaera = Cyathea............................................Cyatheaceae

Gymnosporia = Maytenus ...........................................Celastraceae

**Gymnostoma** L. Johnson..............................................Casuarinaceae

Gymnothrix = Pennisetum ...........................................Poaceae

**Gynandropsis** = Cleome................................................Capparaceae

**Gynoglottis** J.J. Sm. ....................................................Orchidaceae

Gynopogon = Alyxia ...................................................Apocynaceae

**Gynostemma** Blume.....................................................Cucurbitaceae

**Gynotroches** Blume .....................................................Rhizophoraceae

**Gynura** Cass...............................................................Asteraceae

**Gyrinops** Gaertner .....................................................Thymelaeaceae

**Gyrocarpus** Jacq. .......................................................Hernandiaceae

**Habenaria** Willd. .......................................................Orchidaceae

**Habranthus** Herbert ...................................................Amaryllidaceae

**Hackelochloa** Kuntze (Mnesithes) ................................Poaceae

**Haemanthus** L. ...........................................................Amaryllidaceae

**Haemodorum** Sm.........................................................Haemodoraceae

**Halfordia** F. Muell. ....................................................Rutaceae

Halocnemium = Tecticornia ........................................Chenopodiaceae

**Halodule** Endl. ...........................................................Cymodoceaceae

**Halophila** Thouars......................................................Hydrocharitaceae

**Haloragis** Forster & Forster f. ...................................Haloragidaceae

**Halosarcia** P.G. Wilson ..............................................Chenopodiaceae

**Hanguana** Blume ........................................................Hanguanaceae

Hansemannia = Archidendron ....................................... Mimosaceae

Hanslia = Desmodium ................................................ Fabaceae

Haplachne = Dimeria .................................................. Poaceae

Haplochilus = Bulbophyllum........................................ Orchidaceae

Haplochorema see Kaempferia..................................... Zingiberaceae

Haplolobus H.J. Lam. .............................................. Burseraceae

Hardenbergia see Vandasia ....................................... Fabaceae

Hardwickia Roxb. ..................................................... Caesalpiniaceae ........... no record

Harmandia Baillon ................................................... Olacaceae................... no record

Harmsiopanax Warb. ................................................. Araliaceae

Harpullia Roxb. ....................................................... Sapindaceae

Harrisonia Necker = Xeranthemum............................... Asteraceae

Harrisonia R. Br. ex A. Juss....................................... Simaroubaceae

Hartleya Sleumer ..................................................... Icacinaceae

Havilandia = Trigonotis ............................................. Boraginaceae

Hearnia = Aglaia ..................................................... Meliaceae

Hebe Comm. ex Juss. (Veronica, Parahebe) ..................... Scrophulariaceae

Hecatactis = Keysseria .............................................. Asteraceae

Heckelia = Lepianthes............................................... Liliaceae

Heckeria = Pothomorphe............................................ Piperaceae

Hedera see Schefflera................................................ Araliaceae

Hederella = Catanthera.............................................. Melastomataceae

Hedycarya Forster & Forster f. .................................... Monimiaceae

Hedychium J. Koenig................................................. Zingiberaceae

Hedyotis L............................................................... Rubiaceae

Heimerliodendron Skottsb. = Pisonia ............................ Nyctaginaceae

Heleocharis = Eleocharis ............................................ Cyperaceae

Helianthus L. .......................................................... Asteraceae

Helichrysum Miller ................................................... Asteraceae

Helicia Lour. ........................................................... Proteaceae

Heliconia L.............................................................. Heliconiaceae

Heliconiopsis = Heliconia ........................................... Heliconiaceae

Helicteres L. ........................................................... Sterculiaceae

Heliotropium L. ........................................................ Boraginaceae

Hellenia see Alpinia, Costus ....................................... Zingiberaceae

Hellwigia = Alpinia................................................... Zingiberaceae

Helmholtzia F. Muell. ............................................... Philydraceae

Helminthostachys Kaulf. ............................................ Ophioglossaceae

Helothrix = Schoenus ................................................ Cyperaceae

Hemarthria R. Br. .................................................... Poaceae

Hemicyclia = Drypetes ............................................... Euphorbiaceae

Hemiglochidion = Glochidion....................................... Euphorbiaceae

Hemigramma = Tectaria ............................................. Aspleniaceae (Tect.)

**Hemigraphis** Nees (Strobilanthes) ................................. Acanthaceae

Hemigymnia see Panicum ............................................ Poaceae

Hemionitis L. ..................................................... Adiantaceae (Adiant.) ..... no record

Hemipteris = Pteris ............................................... Adiantaceae (Pter.)

Hemitelia = Cyathea ............................................... Cyatheaceae

Henslowia = Dendrotrophe .......................................... Santalaceae

Heptapleurum = Schefflera ......................................... Araliaceae

**Heritiera** Dryander ............................................. Sterculiaceae

**Hernandia** L. .................................................. Hernandiaceae

Herpestis = Bacopa ................................................ Scrophulariaceae

Herzogia see Melicope ............................................. Rutaceae

**Hetaeria** Blume ................................................ Orchidaceae

**Heterogonium** C. Presl. ........................................ Aspleniaceae (Tect.)

Heteroneuron = Bolbitis ........................................... Aspleniaceae (Lom.)

**Heteropogon** Pers. ............................................. Poaceae

Heteropteris see Rhyssopteris ..................................... Malpighiaceae

**Heterospathe** R. Scheffer ...................................... Arecaceae

**Heterostemma** Wight & Arn. ..................................... Asclepiadaceae

**Hevea** Aublet .................................................. Euphorbiaceae

**Hewittia** Wight & Arn. ......................................... Convolvulaceae

**Hibbertia** Andrews ............................................. Dilleniaceae

**Hibiscus** L. ................................................... Malvaceae

Hicriopteris see Dicranopteris, Diplopterygium ................... Gleicheniaceae

Hierochloe see Anthoxanthum ....................................... Poaceae

Himantandra = Galbulimima ......................................... Himantandraceae

**Hippeastrum** Herbert ........................................... Amaryllidaceae

**Hippeophyllum** Schltr. ......................................... Orchidaceae

**Hippocratea** L. ................................................ Celastraceae

**Histiopteris** (J. Agardh) J. Sm. ............................... Dennstaedtiaceae (Denn.)

**Holcosorus** T. Moore ........................................... Polypodiaceae (Micro.)

Holcus see Chrysopogon, Sorghum, Andropogon ...................... Poaceae

**Hollrungia** Schumann ........................................... Passifloraceae

**Holochlamys** Engl. ............................................. Araceae

Hologyne = Coelogyne ............................................. Orchidaceae

**Holostachyum** (Copel.) Ching (Aglaomorpha) ..................... Polypodiaceae (Dryn.)

**Homalanthus** A. Juss. .......................................... Euphorbiaceae

**Homalium** Jacq. ................................................ Flacourtiaceae

Homalocenchrus = Leersia ......................................... Poaceae

**Homalomena** Schott ............................................. Araceae

Hombronia = Pandanus ............................................. Pandanaceae

**Homonoia** Lour. ................................................ Euphorbiaceae

**Hopea** Roxb. ................................................... Dipterocarpaceae

Hormopetalum = Sericolea ......................................... Elaeocarpaceae

Hormopetalum = Sericolea (Elaeocarpaceae) ....................... Rutaceae

**Hornstedtia** Retz. ..................................................... Zingiberaceae
**Horsfieldia** Willd. ................................................. Myristicaceae
**Howeia** Becc. ......................................................... Arecaceae
**Hoya** R. Br. ............................................................ Asclepiadaceae
**Hugonia** L. ............................................................ Linaceae
**Hulemacanthus** S. Moore...................................... Acanthaceae
**Humata** Cav. ......................................................... Davalliaceae (Davall.)
**Hunga** Pancher ex Prance ........................................ Chrysobalanaceae
Hunsteinia see Rapanea (Myrsinaceae)............................ Rutaceae
Hyalosema = Bulbophyllum ......................................... Orchidaceae
**Hybanthus** Jacq. ................................................... Violaceae
**Hydnocarpus** Gaertner ........................................... Flacourtiaceae
**Hydnophytum** Jack.................................................. Rubiaceae
**Hydriastele** H. Wendl. & Drude.................................. Arecaceae
**Hydrilla** Rich. ....................................................... Hydrocharitaceae
**Hydrocharis** L......................................................... Hydrocharitaceae
**Hydrocotyle** L. ...................................................... Apiaceae
**Hydrostemma** Wallich............................................. Nymphaeaceae
**Hygrophila** R. Br. ................................................... Acanthaceae
**Hylomyza** Danser (Dufrenoia) ................................... Santalaceae
**Hylophila** Lindley ................................................... Orchidaceae
**Hymenachne** Pal. .................................................... Poaceae
**Hymenocallis** Salisb. .............................................. Amaryllidaceae
Hymenolepis = Belvisia............................................... Polypodiaceae (Pleo.)
**Hymenophyllum** Sm. ............................................... Hymenophyllaceae
**Hymenorchis** Schltr. ............................................... Orchidaceae
**Hymenosporum** R. Br. ex F. Muell. ............................. Pittosporaceae
**Hyparrhenia** Andersson ex Fourn................................ Poaceae
Hypenanthe (Blume) Blume ......................................... Melastomataceae .......... no record
**Hypericum** L. ........................................................ Clusiaceae
Hypobathrum Blume................................................... Rubiaceae................... no record
**Hypodematium** Kunze ............................................. Aspleniaceae (Athyr.)
**Hypoestes** Sol. ex R. Br.......................................... Acanthaceae
**Hypolepis** Bernh. ................................................... Dennstaedtiaceae (Denn.)
**Hypolytrum** Rich. ................................................... Cyperaceae
**Hypoxis** L. ........................................................... Hypoxidaceae
**Hypserpa** Miers ..................................................... Menispermaceae
**Hyptis** Jacq........................................................... Lamiaceae
**Hyrtanandra** Miq. (Pouzolzia) ................................... Urticaceae
**Ichnanthus** Pal. ..................................................... Poaceae
**Ichnocarpus** R. Br. ................................................. Apocynaceae
Idenburgia = Sphenostemon ........................................ Aquifoliaceae
Ilex L..................................................................... Aquifoliaceae

**Illigera** Blume ............................................................. Hernandiaceae

Illipe = Madhuca ......................................... Sapotaceae

**Ilysanthes** Raf. (Lindernia) ........................................ Scrophulariaceae

**Impatiens** L. ................................................................ Balsaminaceae

**Imperata** Cirillo.......................................................... Poaceae

**Indigofera** L. .............................................................. Fabaceae

**Inga** = Cathormion ......................................... Mimosaceae

**Inocarpus** Forster & Forster f. ..................................... Fabaceae

**Intsia** Thouars ............................................................ Caesalpiniaceae

**Ioedes** Blume.............................................................. Icacinaceae

Ionidium = Hybanthus ................................... Violaceae

**Iphigenia** Kunth .......................................................... Liliaceae

**Ipomoea** L. ................................................................. Convolvulaceae

**Iresine** P. Browne........................................................ Amaranthaceae

Irina = Pometia ............................................. Sapindaceae

**Isachne** R. Br. ............................................................. Poaceae

Isanthera = Rhynchotechum ........................... Gesneriaceae

**Ischaemum** L............................................................... Poaceae

**Ischnea** F. Muell.......................................................... Asteraceae

**Ischnocentrum** Schltr. ................................................. Orchidaceae

**Ischnostemma** King & Gamble ..................................... Asclepiadaceae

**Isoetes** L. ................................................................... Isoetaceae

Isolepis = Scirpus ......................................... Cyperaceae

Isoloma see Lindsaea ..................................... Dennstaedtiaceae (Lind.)

Isomerocarpa = Dryadodaphne ...................... Monimiaceae

Isotoma = Solenopsis ..................................... Campanulaceae

Ithyocaulon = Orthiopteris.............................. Dennstaedtiaceae (Denn.)

**Itoa** Hemsley............................................................... Flacourtiaceae

**Ixeris** (Cass.) Cass. (Lactuca) ...................................... Asteraceae

**Ixonanthes** Exell & Mendonça...................................... Ixonanthaceae

**Ixora** L. ...................................................................... Rubiaceae

**Jacaranda** Juss. ........................................................... Bignoniaceae

**Jacquemontia** Choisy ................................................... Convolvulaceae

**Jadunia** Lindau............................................................ Acanthaceae

**Jagera** Blume .............................................................. Sapindaceae

Jambosa = Syzygium ..................................... Myrtaceae

**Jasminum** L................................................................. Oleaceae

Jatropha see Aleurites, Manihot ...................... Euphorbiaceae

Jeanneretia = Pandanus.................................. Pandanaceae

**Joinvillea** Gaudich. ex Brongn. & Gris............................ Joinvilleaceae

**Josephinia** Vent. .......................................................... Pedaliaceae

Jossinia = Eugenia ......................................... Myrtaceae

**Juncus** L. ................................................................... Juncaceae

Jungia L. f. (Trinacte).................................... Asteraceae.................. no record

Jussiaea = Ludwigia ................................................. Onagraceae

**Justicia** L. ................................................. Acanthaceae

**Kaempferia** L. ................................................. Zingiberaceae

Kaernbachia = Turpinia ................................................. Staphyleaceae

**Kairoa** Philipson ................................................. Monimiaceae

**Kairothamnus** Airy Shaw ................................................. Euphorbiaceae

Kajewskia see Veitchia ................................................. Arecaceae

**Kajewskiella** Merr. & Perry ................................................. Rubiaceae

**Kalanchoe** Adans. ................................................. Crassulaceae

Kalappia Kosterm. ................................................. Caesalpiniaceae ............ no record

**Kandelia** (DC.) Wight & Arn. ................................................. Rhizophoraceae

**Kania** Schltr. (Metrosideros) ................................................. Myrtaceae

Katharinea = Epigeneium ................................................. Orchidaceae

**Kayea** Wall (Mesua) ................................................. Clusiaceae

**Kedrostis** Medikus ................................................. Cucurbitaceae

**Kelleria** Endl. (Drapetes) ................................................. Thymelaeaceae

Kennedia Vent. ................................................. Fabaceae .................... no record

Kennedya (Kennedia) = Vandasina ................................................. Fabaceae

Kentia = Gronophyllum ................................................. Arecaceae

**Kentrochrosia** Lauterb. & K. Schum. ................................................. Apocynaceae

**Keraudrenia** Gay ................................................. Sterculiaceae

**Keysseria** Lauterb. ................................................. Asteraceae

**Khaya** A. Juss. ................................................. Meliaceae

**Kibara** Endl. ................................................. Monimiaceae

Kibessia = Pternandra ................................................. Melastomataceae

Kiesera = Tephrosia ................................................. Fabaceae

**Kingiodendron** Harms (Oxystigma) ................................................. Caesalpiniaceae

Kissodendron = Polyscias ................................................. Araliaceae

**Kjellbergiodendron** Burret ................................................. Myrtaceae

**Kleinhovia** L. ................................................. Sterculiaceae

**Knema** Lour. ................................................. Myristicaceae

**Knoxia** L. ................................................. Rubiaceae

**Koompassia** Maingay ................................................. Caesalpiniaceae

**Koordersiodendron** Engl. ................................................. Anacardiaceae

Kopsia see Kentrochrosia ................................................. Apocynaceae

**Korthalsella** Tieghem ................................................. Viscaceae

**Korthalsia** Blume ................................................. Arecaceae

Krausella = Pouteria ................................................. Sapotaceae

**Kuhlhasseltia** J.J. Sm. ................................................. Orchidaceae

**Kunstleria** Prain ................................................. Fabaceae

Kurrimia = Bhesa ................................................. Celastraceae

Kyllinga = Cyperus ................................................. Cyperaceae

**Labisia** Lindley ................................................. Myrsinaceae

**Lablab** Adans. (Dolichos)...............................................Fabaceae

**Laccospadix** Drude & H. Wendl. (Calyptrocalyx) ..............Arecaceae

Lactaria = Ochrosia .................................................Apocynaceae

**Lactuca** L. ............................................................Asteraceae

**Lagenaria** Ser. ..................................................Cucurbitaceae

Lagenifera = Lagenophora...........................................Asteraceae

**Lagenophora** Cass.....................................................Asteraceae

**Lagerstroemia** L.......................................................Lythraceae

Laguncularia see Lumnitzera.......................................Combretaceae

Lagurus see Imperata .....................................................Poaceae

**Lamechites** Markgraf = Micrechtites ...........................Apocynaceae

**Lamiodendron** Steenis (Fernandoa)...............................Bignoniaceae

Lamiofrutex = Vavaea (Meliaceae) ...............................Rutaceae

Lampocarya = Gahnia ...............................................Cyperaceae

**Langsdorffia** C. Martius ........................................Balanophoraceae

Lannea A. Rich. .......................................Anacardiaceae ............no record

**Lansium** Corr. Serr. ..................................................Meliaceae

**Lantana** L. .............................................................Verbenaceae

Laplacea = Gordonia....................................................Theaceae

**Laportea** Gaudich.....................................................Urticaceae

**Lasia** Lour.................................................................Araceae

Lasianthera see Gonocaryum, Medusanthera ....................Icacinaceae

**Lasianthus** Jack ......................................................Rubiaceae

**Lasiobema** (Korth.) Miq. (Bauhinia)............................Caesalpiniaceae

Lasiostoma see Strychnos (Loganiaceae) .......................Rubiaceae

Lastrea = Thelypteris ..............................................Thelypteridaceae

**Lastreopsis** Ching....................................................Aspleniaceae (Tect.)

**Lathyrus** L..............................................................Fabaceae

Latouria = Dendrobium ...............................................Orchidaceae

Laurentia = Solenopsis ...............................................Campanulaceae

**Lauterbachia** Perkins .............................................Monimiaceae

**Lawsonia** L. ............................................................Lythraceae

**Lecanopteris** Reinw. ................................................Polypodiaceae (Micro.)

**Lecanorchis** Blume...................................................Orchidaceae

**Lechenaultia** R. Br....................................................Goodeniaceae

Lectandra = Poaephyllum ...........................................Orchidaceae

**Leea** Royen ex L. .....................................................Leeaceae

**Leersia** Sw. ..............................................................Poaceae

**Legnephora** Miers .....................................................Menispermaceae

Leiospermum = Psilotrichum .......................................Amaranthaceae

**Lemmaphyllum** C. Presl............................................Polypodiaceae (Pleo.)

**Lemna** L. ...............................................................Lemnaceae

**Lepeostegeres** Blume...............................................Loranthaceae

**Lepianthes** Raf. (Piper) ...........................................Piperaceae

**Lepidagathis** Willd. ................................................... Acanthaceae
**Lepiderema** Radlk. ................................................... Sapindaceae
**Lepidium** L. ................................................... Brassicaceae
Lepidocaulon see Histiopteris ....................................... Dennstaedtiaceae (Denn.)
**Lepidogyne** Blume................................................... Orchidaceae
**Lepidopetalum** Blume ................................................... Sapindaceae
**Lepidosperma** Labill. ................................................... Cyperaceae
**Lepinia** Decne. ................................................... Apocynaceae
**Lepiniopsis** Valeton ................................................... Apocynaceae
**Lepionurus** Blume................................................... Opiliaceae
**Lepironia** Pers. ................................................... Cyperaceae
**Lepisanthes** Blume ................................................... Sapindaceae
Lepisorus see Crypsinus, Microsorum ........................... Polypodiaceae (Micro.)
**Lepistemon** Blume................................................... Convolvulaceae
**Leptaspis** R. Br. ................................................... Poaceae
**Leptocarpus** R. Br. ................................................... Restionaceae
**Leptochilus** Kaulf. ................................................... Polypodiaceae (Micro.)
**Leptochloa** Pal. ................................................... Poaceae
Leptogramma = Thelypteris ....................................... Thelypteridaceae
**Leptolepia** Prantl (Microlepia) ...................................... Dennstaedtiaceae (Denn.)
**Leptonychia** Turcz. ................................................... Sterculiaceae
Leptophoenix = Nengella ...................................... Arecaceae
**Leptopteris** C. Presl................................................... Osmundaceae
**Leptopus** Decne. ................................................... Euphorbiaceae
Leptosiphonium F. Muell. = Ruellia ........................... Acanthaceae
**Leptospermum** Forster & Forster f. ........................... Myrtaceae
**Lepturus** R. Br. ................................................... Poaceae
Leschenaultia = Lechenaultia ...................................... Goodeniaceae
**Lespedeza** Michaux................................................... Fabaceae
**Leucaena** Benth. ................................................... Mimosaceae
**Leucas** R. Br. ................................................... Lamiaceae
Leucocorema = Trichadenia ...................................... Flacourtiaceae
**Leuconotis** Jack ................................................... Apocynaceae
**Leucopogon** R. Br. (Styphelia)................................... Epacridaceae
**Leucostegia** C. Presl ................................................... Davalliaceae (Davall.)
**Leucosyke** Zoll. & Moritzi ...................................... Urticaceae
**Levieria** Becc. ................................................... Monimiaceae
**Libertia** Sprengel (Sisyrinchum)................................... Iridaceae
**Libocedrus** Endl. ................................................... Cupressaceae
**Licania** Aublet................................................... Chrysobalanaceae
**Licuala** Thunb. ................................................... Arecaceae
**Ligustrum** L. ................................................... Oleaceae
**Limacia** Lour. ................................................... Menispermaceae

Limnanthemum = Nymphoides (Menyanthaceae) ............... Gentianaceae

**Limnophila** R. Br. ..................................................... Scrophulariaceae

**Limnophyton** Miq. ................................................... Alismataceae

**Lindenbergia** Lehm. ................................................. Scrophulariaceae .......... no record

**Lindera** Thunb. ...................................................... Lauraceae

**Lindernia** All. ......................................................... Scrophulariaceae

**Lindsaea** Dryander ex Sm. ....................................... Dennstaedtiaceae (Lind.)

**Lindsayomyrtus** B. Hyland & Steenis ........................... Myrtaceae

Linociera = Chionanthus ............................................ Oleaceae

Linodorum = Epipactis ................................................ Orchidaceae

**Linospadix** H. Wenl. ................................................ Arecaceae

**Liparis** Rich. ........................................................... Orchidaceae

**Lipocarpha** R. Br. .................................................... Cyperaceae

Lippia see Phyla ....................................................... Verbenaceae

**Liquidambar** L. (Altingia) ........................................ Hamamelidaceae

**Litchi** Sonn. ........................................................... Sapindaceae

**Lithocarpus** Blume .................................................. Fagaceae

**Lithospermum** L. (Trigonotis) ................................... Boraginaceae

**Litosanthes** Blume .................................................. Rubiaceae

**Litsea** Lam. ........................................................... Lauraceae

**Livistona** R. Br. ...................................................... Arecaceae

**Lobelia** L. .............................................................. Lobeliaceae

Lobogyne = Appendicula ............................................ Orchidaceae

Lochnera = Catharanthus ............................................ Apocynaceae

**Loesneriella** A.C. Sm. .............................................. Celastraceae

**Loheria** Merr. ........................................................ Myrsinaceae

Lolanara = Mammea ................................................... Clusiaceae

Lolium L. ................................................................. Poaceae ..................... no record

**Lomagramma** J. Sm. ................................................ Aspleniaceae (Lom.)

**Lomandra** Labill. ..................................................... Xanthorrhoeaceae

Lomaria = Blechnum .................................................. Blechnaceae

**Lomariopsis** Fée ..................................................... Aspleniaceae (Lom.)

**Lonchocarpus** Kunth ............................................... Fabaceae

Longetia = see Kairothamnus ...................................... Euphorbiaceae

**Lonicera** L. ............................................................ Caprifoliaceae

Lophatherum see Centotheca ....................................... Poaceae

Lophidium = Schizaea ................................................ Schizaeaceae

**Lophopetalum** Wight ex Arn. ..................................... Celastraceae

**Lophopyxis** Hook f. ................................................. Celastraceae

**Lophoschoenus** Stapf (Costularia) .............................. Cyperaceae

**Loranthus** Jacq. ...................................................... Loranthaceae

**Lotononis** (DC.) Ecklon & Zeyher .............................. Fabaceae

**Lotus** L. ................................................................ Fabaceae

Lourea = Christia ....................................................... Fabaceae

**Loxogramme** (Blume) C. Presl ..................................... Grammitidaceae (Loxo.)

**Loxoscaphe** T. Moore (Asplenium) ............................... Aspleniaceae (Asplen.)

Loxotis = Rhynchoglossum.......................................... Gesneriaceae

**Lucinaea** DC...................................................... Rubiaceae

Lucuma = Pouteria................................................. Sapotaceae

**Ludwigia** L...................................................... Onagraceae

Luerssenia Kuhn ex Luerssen (Tectaria) ......................... Aspleniaceae (Tect.) ...... no record

**Luffa** Miller.................................................... Cucurbitaceae

**Luisia** Gaudich. ............................................... Orchidaceae

**Lumnitzera** Willd............................................... Combretaceae

**Lunasia** Blanco................................................. Rutaceae

Lunathyrium = Diplazium .......................................... Aspleniaceae (Athyr.)

**Lupinus** L. ..................................................... Fabaceae

**Luvunga** Buch.-Ham. ex Wight & Arn.......................... Rutaceae

**Luzula** DC. .................................................... Juncaceae

**Luzuriaga** Ruíz & Pavón......................................... Smilacaceae

Lycianthes see Solanum............................................ Solanaceae

**Lycopersicon** Miller (Solanum) ................................. Solanaceae

**Lycopodium** L. ................................................. Lycopodiaceae

**Lygodium** Sw.................................................... Schizaeaceae

**Lyonsia** R. Br. (Parsonsia)..................................... Apocynaceae

**Lysimachia** L. ................................................. Primulaceae

**Lysiphyllum** (Benth.) De Wit (Bauhinia) ....................... Caesalpiniaceae

**Lythrum** L. .................................................... Lythraceae

Maba = Diospyros.................................................. Ebenaceae

**Macadamia** F. Muell. ........................................... Proteaceae

**Macaranga** Thouars.............................................. Euphorbiaceae

**Machaerina** Vahl ............................................... Cyperaceae

**Mackinlaya** F. Muell............................................ Araliaceae

**Maclura** Nutt. ................................................. Moraceae

**Macodes** (Blume) Lindley ....................................... Orchidaceae

**Macrococculus** Becc............................................. Menispermaceae

Macroglena = Selenodesmium ....................................... Hymenophyllaceae

**Macroglossum** Copel. ........................................... Marattiaceae ............... no record

**Macrolenes** Naudin ex. Miq..................................... Melastomataceae

Macrolobium see Intsia............................................ Caesalpiniaceae

**Macropiper** Miq. (Piper) ....................................... Piperaceae

**Macropsychanthus** Harms ex Schumann & Lauterb............. Fabaceae

**Macroptilium** (Benth.) Urban.................................... Fabaceae

**Macrosolen** (Blume) Blume....................................... Loranthaceae

**Macrothelypteris** (H. Itô) Ching ............................... Thelypteridaceae

**Macrotyloma** (Wight & Arn.) Verdc. ............................ Fabaceae

Macrozamia Miq. .................................................. Zamiaceae................... no record

Macrozanonia = Alsomitra .......................................... Cucurbitaceae

**Madhuca** J. Gmelin ................................................. Sapotaceae

**Maesa** Forssk. ....................................................... Myrsinaceae

**Maesopsis** Engl. .................................................... Rhamnaceae

**Magnolia** L........................................................... Magnoliaceae

**Magodendron** Vink ................................................. Sapotaceae

**Malachra** L. ......................................................... Malvaceae

**Malaisia** Blanco .................................................... Moraceae

**Malaxis** Sol. ex Sw................................................. Orchidaceae

**Malleola** J.J. Sm. & Schltr........................................ Orchidaceae

**Mallotus** Lour. ...................................................... Euphorbiaceae

**Malpighia** L. ........................................................ Malpighiaceae

**Malvastrum** A. Gray ............................................... Malvaceae

**Mammea** L............................................................ Clusiaceae

**Mangifera** L.......................................................... Anacardiaceae

**Manihot** Miller ...................................................... Euphorbiaceae

**Manilkara** Adans. .................................................. Sapotaceae

**Maniltoa** R. Scheffer .............................................. Caesalpiniaceae

Manisuris see Rottboellia .......................................... Poaceae

**Maoutia** Wedd. ..................................................... Urticaceae

**Mapania** Aublet..................................................... Cyperaceae

Mapouria Aublet ..................................................... Rubiaceae................... no record

Mappa = Macaranga ................................................. Euphorbiaceae

**Maranta** L. ........................................................... Marantaceae

Maranthes see Parinari ............................................. Chrysobalanaceae

**Marantochloa** Brongn. ex Gris .................................. Marantaceae

**Marattia** Sw.......................................................... Marattiaceae

**Margaritaria** L. f. .................................................. Euphorbiaceae

Mariscus = Cyperus .................................................. Cyperaceae

**Marsdenia** R. Br...................................................... Asclepiadaceae

**Marsilea** L. ........................................................... Marsileaceae

**Martynia** L............................................................ Pedaliaceae

Marumia = Macrolenes .............................................. Melastomataceae

**Maschalodesme** Schumann & Lauterb. ......................... Rubiaceae

Massoia = Cryptocarya.............................................. Lauraceae

**Mastersia** Benth...................................................... Fabaceae .................... no record

**Mastixia** Blume...................................................... Cornaceae

**Mastixiodendron** Melchior ....................................... Rubiaceae

Matonia R. Br.......................................................... Matoniaceae................... no record

Matthaea Blume....................................................... Monimiaceae............... no record

Maughania J. St-Hil. ................................................ Fabaceae .................... no record

Maurandya = Asarina................................................ Scrophulariaceae

Maxillaria Ruíz & Pavón ........................................... Orchidaceae................ no record

**Maytenus** Molina ................................................... Celastraceae

Mazus Lour. ............................................... Scrophulariaceae

Mearnsia = Metrosideros ........................................ Myrtaceae

Mecodium C. Presl ex Copel. (Hymenophyllum) .............. Hymenophyllaceae

Medicago L. ............................................... Fabaceae

Medinilla Gaudich. ....................................... Melastomataceae

Mediocalcar J.J. Sm. ..................................... Orchidaceae

Medicosma Hook. f. ....................................... Rutaceae .................... no record

Medusanthera Seemann ..................................... Icacinaceae

Meiogyne Miq. ............................................ Annonaceae

Melaleuca L. ............................................. Myrtaceae

Melanococca = Rhus (Anacardiaceae) ........................... Rutaceae

Melanolepis Reichb. f. ex Zoll. ........................... Euphorbiaceae

Melanthera J.P. Rohr ..................................... Asteraceae

Melastoma L. ............................................. Melastomataceae

Melhania Forssk. ......................................... Sterculiaceae

Melia L. ................................................. Meliaceae

Melicope Forster & Forster f. ............................ Rutaceae

Melinis Pal. ............................................. Poaceae

Melio-Schinzia = Chisocheton ............................. Meliaceae

Meliosma Blume ........................................... Sabiaceae

Mella = Bacopa ........................................... Scrophulariaceae

Melochia L. .............................................. Sterculiaceae

Melodinus Forster & Forster f. ........................... Apocynaceae

Melodorum Lour. .......................................... Annonaceae

Melothria L. ............................................. Cucurbitaceae

Memecylon L. ............................................. Melastomataceae

Memorialis = Hyrtanandra. ................................ Urticaceae

Mentha L. ................................................ Lamiaceae

Meringium C. Presl (Hymenophyllum) ....................... Hymenophyllaceae

Merinthosorus Copel. ..................................... Polypodiaceae (Dryn.)

Merope M. Roemer (Atalantia) ............................. Rutaceae

Merremia Dennst. ex Endl. ................................ Convolvulaceae

Merrilliodendron Kaneh. .................................. Icacinaceae

Meryta Forster & Forster f. .............................. Araliaceae

Mesochlaena = Sphaerostephanos ........................... Thelypteridaceae

Mesona Blume. ............................................ Lamiaceae

Mesophlebion Holttum ..................................... Thelypteridaceae

Messerschmidia = Argusia. ................................ Boraginaceae

Mesua L. ................................................. Clusiaceae

Metabolos Blume (Hedyotis) ............................... Rubiaceae

Metadina Bakh. f. ........................................ Rubiaceae

Metrosideros Banks ex Gaertner ........................... Myrtaceae

Metroxylon Rottb. ........................................ Arecaceae

Mezonevron Desf. (Caesalpinia) ..................................... Caesalpiniaceae
Michelia L............................................................................. Magnoliaceae
Micrechites Miq..................................................................... Apocynaceae
Microcitrus Swingle................................................................ Rutaceae
Microcos L. ........................................................................... Tiliaceae
Microglossa DC. .................................................................. Asteraceae
Microgonium C. Presl ......................................................... Hymenophyllaceae
Microlaena = Ehrharta .............................................. Poaceae
Microlepia C. Presl .............................................................. Dennstaedtiaceae (Denn.)
Micromelum Blume ............................................................. Rutaceae
Micropera Lindley................................................................. Orchidaceae
Microsorum Link (Microsorium, Polypodium).................. Polypodiaceae (Micro.)
Microstegium Nees................................................................ Poaceae
Microstylis = Malaxis ........................................................ Orchidaceae
Microtatorchis Schltr. .......................................................... Orchidaceae
Microtis R. Br. ...................................................................... Orchidaceae............... no record
Microtoena Prain ................................................................. Lamiaceae
Microtrichomanes (Mett.) Copel. ....................................... Hymenophyllaceae
Microtropis Wallich ex Meissner ....................................... Celastraceae................ no record
Mikania Willd....................................................................... Asteraceae
Milicia Sim .......................................................................... Moraceae
Milium see Isachne, Eriochloa............................................ Poaceae
Miliusa Leschen ex A. DC. ................................................ Annonaceae
Millettia Wight & Arn.......................................................... Fabaceae
Millingtonia see Meliosma .......................................... Sabiaceae
Miltonia Lindley ................................................................... Orchidaceae
Mimosa L. ............................................................................. Mimosaceae
Mimulus see Torenia ................................................... Scrophulariaceae
Mimusops L. ......................................................................... Sapotaceae
Mina Llave & Lex................................................................ Convolvulaceae
Mirabilis L............................................................................. Nyctaginaceae
Miscanthus Andersson ................................................. Poaceae
Mischobulbum = Tainia......................................................... Orchidaceae
Mischocarpus Blume ........................................................... Sapindaceae
Mischocodon = Mischocarpus ............................................. Sapindaceae
Mischopleura see Sericolea................................................ Elaeocarpaceae
Missiessya = Leucosyke....................................................... Urticaceae
Mitracarpus Zucc. (Mitracarpum) .............................. Rubiaceae
Mitragyna Korth.................................................................... Rubiaceae
Mitrasacme Labill. ............................................................... Loganiaceae
Mitrastemma Makino (Mitrastemmon) ........................... Mitrastemmataceae
Mitrella = Fissistigma ........................................................ Annonaceae
Mitreola L.............................................................................. Loganiaceae
Mitrephora (Blume) Hook. f. & Thomson ...................... Annonaceae

Modecca = Adenia ..................................................... Passifloraceae
Moerenhoutia = Malaxis ............................................. Orchidaceae
Moghania = Flemingia ............................................... Fabaceae
**Mollinedia** Ruíz & Pavón (Wilkiea)............................... Monimiaceae
**Mollugo** L. ........................................................ Molluginaceae
**Momordica** L....................................................... Cucurbitaceae
**Monachosorum** Hance .............................................. Dennstaedtiaceae (Monach.)
**Monanthocitrus** Tanaka ........................................... Rutaceae
Monarthrocarpus = Desmodium .................................... Fabaceae
**Monochoria** C. Presl............................................... Pontederiaceae
**Monogramma** Comm. ex Schkuhr ................................ Adiantaceae (Vitt.)
**Monophyllaea** R. Br. .............................................. Gesneriaceae
Monosepalum = Bulbophyllum .................................... Orchidaceae
**Monostachya** Merr. (Danthonia) ................................. Poaceae
Monstera see Rhaphidophora....................................... Araceae
**Montia** L............................................................ Portulacaceae
**Morinda** L. ........................................................ Rubiaceae
**Moringa** Adans..................................................... Moringaceae
Morolobium = Archidendron ....................................... Mimosaceae
Morus see Streblus .................................................. Moraceae
**Moschosma** Reichb. (Basilicum)................................... Lamiaceae
**Mucuna** Adans. .................................................... Fabaceae
**Muehlenbeckia** Meissner............................................ Polygonaceae
Muellerina see Cecarria ............................................ Loranthaceae
**Muhlenbergia** Schreber (Muehlenbergia) ........................ Poaceae
**Mukia** Arn. ........................................................ Cucurbitaceae
**Mundulea** (DC.) Benth. ............................................ Fabaceae
**Murdannia** Royle .................................................. Commelinaceae
**Murraya** Koenig ex L. ............................................. Rutaceae
**Musa** L............................................................. Musaceae
**Mussaenda** L. ...................................................... Rubiaceae
**Mycetia** Reinw...................................................... Rubiaceae
**Myoporum** Banks & Sol. ex Forster f. ........................... Myoporaceae
**Myosotis** L. ........................................................ Boraginaceae
**Myriactis** Less. ..................................................... Asteraceae
**Myrica** L............................................................ Myricaceae
**Myriodon** (Copel.) Copel. .......................................... Hymenophyllaceae
**Myriophyllum** L. ................................................... Haloragidaceae
**Myristica** Gronov................................................... Myristicaceae
**Myrmecodia** Jack .................................................. Rubiaceae
Myrmecophila = Lecanopteris ..................................... Polypodiaceae (Micro.)
**Myrmedoma** = Myrmephytum ..................................... Rubiaceae
**Myrmephytum** Becc. ............................................... Rubiaceae

Myrsine L. (Rapanea) ............................................... Myrsinaceae

Myrtella F. Muell. .................................................. Myrtaceae

Myrtus L. ........................................................... Myrtaceae

Myxopyrum Blume .................................................... Oleaceae

Nageia see Decussocarpus ........................................... Podocarpaceae

Najas L. ............................................................ Najadaceae

Nani see Xanthostemon .............................................. Myrtaceae

Nanochilus Schumann ................................................ Zingiberaceae ............. no record

Nardus see Eremochloa .............................................. Poaceae

Nasturtium R. Br. (Rorippa) ........................................ Brassicaceae

Nastus Juss. ........................................................ Poaceae

Nauclea Merr. ....................................................... Rubiaceae

Naumannia = Riedelia ............................................... Zingiberaceae

Neisosperma Raf. (Ochrosia) ........................................ Apocynaceae

Nelitris = Decaspermum ............................................. Myrtaceae

Nelsonia R. Br. ..................................................... Acanthaceae ............... no record

Nelumbo Adans. (Nelumbium) ......................................... Nelumbonaceae

Nematopteris Alderw. ............................................... Grammitidaceae ........... no record

Nenga H. Wendl. & Drude ............................................ Arecaceae

Nengella Becc. (Gronophyllum) ...................................... Arecaceae

Neoalsomitra Hutch. ................................................ Cucurbitaceae

Neocheiropteris Christ ............................................. Polypodiaceae (Pleo.) ..... no record

Neodissochaeta Bakh. f. (Dissochaeta) .............................. Melastomataceae

Neodistemon Babu & A.N. Henry ...................................... Urticaceae

Neojunguhnia = Vaccinium ........................................... Ericaceae

Neolitsea (Benth.) Merr. ........................................... Lauraceae

Neomphalea see Omphalea ............................................ Euphorbiaceae

Neonauclea Merr. ................................................... Rubiaceae

Neonotonia Lackey .................................................. Fabaceae

Neoscortechinia Pax ................................................ Euphorbiaceae

Neosepicaea Diels .................................................. Bignoniaceae

Neowollastonia = Melodinus ......................................... Apocynaceae

Nepenthes L. ....................................................... Nepenthaceae

Nephelaphyllum see Collabium ....................................... Orchidaceae

Nephelium L. ....................................................... Sapindaceae

Nephrodium = Dryopteris ............................................ Aspleniaceae (Dryopt.)

Nephrolepis Schott ................................................. Davalliaceae (Oleand.)

Neptunia Lour. ..................................................... Mimosaceae

Nerium L. .......................................................... Apocynaceae

Nertera Banks & Sol. ex Gaertner ................................... Rubiaceae

Nervilia Comm. ex Gaudich. ......................................... Orchidaceae

Nesopteris Copel. (Trichomanes) .................................... Hymenophyllaceae

Neuburgia Blume .................................................... Loganiaceae

Neurachne see Sacciolepis .......................................... Poaceae

Neurogramma = Gymnopteris .................................... Adiantaceae (Adiant.) ..... no record
Neustanthus = Pueraria............................................... Fabaceae
Neuwiedia Blume ...................................................... Orchidaceae
Nicolaia Horan. (Etlingera) ......................................... Zingiberaceae
Nicotiana L. ............................................................. Solanaceae
Niemeyera see Chrysophyllum and Dysoxylum (Meliaceae) ... Sapotaceae
Nipa = Nypa............................................................. Arecaceae
Niphobolus = Pyrrosia ................................................ Polypodiaceae (Platy.)
Normanbya F. Muell. ex Becc. (Ptychosperma) ................ Arecaceae
Nortuia Hook. f. ........................................................ Sapotaceae
Notanthera (DC.) G. Don f. (Loranthus) ......................... Loranthaceae
Nothaphoebe Blume.................................................... Lauraceae
Nothocnide Blume ...................................................... Urticaceae
Nothofagus Blume...................................................... Fagaceae
Notholaena R. Br. (Cheilanthes) ................................... Adiantaceae (Adiant.) .... no record
Nothopanax = Polyscias............................................... Araliaceae
Nothopegiopsis = Semecarpus ...................................... Anacardiaceae
Nothoruellia Bremek. & Nannenga-Bremek. = Ruellia ........ Acanthaceae
Notothixos Oliver....................................................... Viscaceae
Nouhuysia = Sphenostemon .......................................... Aquifoliaceae
Nyctocalos Teijsm. & Binnend. ..................................... Bignoniaceae................ no record
Nymania see Phyllanthus (Euphorbiaceae) ....................... Meliaceae
Nymphaea L............................................................... Nymphaeaceae
Nymphoides Hill ........................................................ Menyanthaceae
Nypa Steck (Nipa) ..................................................... Arecaceae
Nyssa L. .................................................................. Nyssaceae
Oberonia Lindley ....................................................... Orchidaceae
Ochrocarpus = Mammea .............................................. Clusiaceae
Ochroma SW. ............................................................ Bombacaceae
Ochrosia Juss............................................................ Apocynaceae
Ochtocharis Blume ..................................................... Melastomataceae
Ocimum L................................................................. Lamiaceae
Octamyrtus Diels ....................................................... Myrtaceae
Octarrhena Thwaites.................................................... Orchidaceae
Octomeles Miq. ......................................................... Datiscaceae
Octospermum Airy Shaw .............................................. Euphorbiaceae
Odina = Lannea.......................................................... Anacardiaceae ............ no record
Odontosoria Fée ......................................................... Dennstaedtiaceae (Lind.)
Oenanthe L. .............................................................. Apiaceae
Oenotrichia Copel. (Microlepia)..................................... Dennstaedtiaceae (Denn.)
Olax L...................................................................... Olacaceae
Oldenlandia L. (Hedyotis) ............................................ Rubiaceae
Olea L...................................................................... Oleaceae

Oleandra Cav. ........................................................ Davalliaceae (Oleand.)

Oleandropsis Copel. ............................................... Polypodiaceae (Micro.)

Olearia Moench ..................................................... Asteraceae

Omphalea L. ........................................................... Euphorbiaceae

Omphalopus see Dissochaeta ................................. Melastomataceae

Oncocarpus = Semecarpus ..................................... Anacardiaceae

Oncodostigma Diels ............................................... Annonaceae

Oncosperma Blume ................................................ Arecaceae ................... no record

Onychium Blume = Dendrobium .............................. Orchidaceae

Onychium Kaulf. .................................................... Adiantaceae (Adiant.)

Operculina A. Silva Manso ..................................... Convolvulaceae

Ophioderma = Ophioglossum ................................. Ophioglossaceae

Ophioglossum L. ................................................... Ophioglossaceae

Ophiorrhiza L. ....................................................... Rubiaceae

Ophiuros Gaertner f. .............................................. Poaceae

Opilia Roxb. .......................................................... Opiliaceae

Oplismenus Pal. ..................................................... Poaceae

Opocunonia = Caldcluvia ........................................ Cunoniaceae

Orania Zipp. ........................................................... Arecaceae

Orchiodes = Goodyera ............................................ Orchidaceae

Orchipeda = Voacanga ........................................... Apocynaceae

Oreiostachys see Nastus ......................................... Poaceae

Oreobolus R. Br. .................................................... Cyperaceae

Oreocallis R. Br. .................................................... Proteaceae

Oreocnide Miq. (Villebrunea) ................................. Urticaceae

Oreogrammitis = Grammitis .................................... Grammitidaceae

Oreomitra Diels ..................................................... Annonaceae

Oreomyrrhis Endl. ................................................. Apiaceae

Oreothyrsus Lindau (Ptyssiglottis) .......................... Acanthaceae

Ormocarpum Pal. ................................................... Fabaceae

Ormosia Jackson .................................................... Fabaceae

Ornithocephalochloa = Thuarea .............................. Poaceae

Ornithochilus (Lindley) Wallich ex Benth. ............... Orchidaceae

Orophea Blume ...................................................... Annonaceae

Orsidice = Thrixspermum ....................................... Orchidaceae

Orthiopteris Copel. (Saccoloma) ............................. Dennstaedtiaceae (Denn.)

Ortholobium = Archidendron ................................. Mimosaceae

Orthosiphon Benth. ............................................... Lamiaceae

Oryza L. ................................................................ Poaceae

Osbeckia L. ........................................................... Melastomataceae

Osbornia F. Muell. ................................................ Myrtaceae

Osmelia Thwaites ................................................... Flacourtiaceae

Osmoxylon Miq. .................................................... Araliaceae

Osmunda see Helminthostachys (Ophioglossaceae) ..... Osmundaceae

Ostyocarpus Hook. f. ....................Fabaceae ....................no record
Osyricera = Bulbophyllum...........................Orchidaceae
Otanthera Blume .....................................Melastomataceae
Ottelia Pers. ..........................................Hydrocharitaceae
Ottochloa Dandy ....................................Poaceae
Ourouparia = Uncaria ..............................Rubiaceae
Outea see Intsia.......................................Caesalpiniaceae ............no record
Oxalis L..................................................Oxalidaceae
Oxyanthera = Thelasis...............................Orchidaceae
Oxychlamys Schltr. = Aeschynanthus .............Gesneriaceae
Oxymitra = Friesodielsia............................Annonaceae
Oxyrhynchus Brandegee .............................Fabaceae
Oxyspora DC............................................Melastomataceae
Oxystophyllum = Dendrobium.....................Orchidaceae
Oxytenanthera Munro ...............................Poaceae
Pachira Aublet ........................................Bombacaceae
Pachycentria Blume .................................Melastomataceae
Pachygone Miers .....................................Menispermaceae
Pachyne = Phaius.....................................Orchidaceae
Pachyrhizus Rich. ex DC. ..........................Fabaceae
Pachystoma Blume ...................................Orchidaceae
Pachystylus Schumann ..............................Rubiaceae
Paederia L...............................................Rubiaceae
Paesia J. St-Hil.........................................Dennstaedtiaceae (Denn.)
Pagiantha = Tabernaemontana ....................Apocynaceae
Pahudia = Intsia.......................................Caesalpiniaceae
Palaquium Blanco ....................................Sapotaceae
Palmeria F. Muell. ...................................Monimiaceae
Palmervandenbroekia = Polyscias.................Araliaceae
Panax see Polyscias ..................................Araliaceae
Pancratium L. .........................................Amaryllidaceae
Pandanus Parkinson .................................Pandanaceae
Pandorea (Engl.) Spach .............................Bignoniaceae
Pangium Reinw. .......................................Flacourtiaceae
Panicum L. .............................................Poaceae
Paphia = Agapetes....................................Ericaceae
Paphiopedilum Pfitzer ...............................Orchidaceae
Papilionopsis = Desmodium .......................Fabaceae
Papuacalia Veldk......................................Asteraceae
Papuacedrus = Libocedrus..........................Cupressaceae
Papuaea Schltr. .......................................Orchidaceae
Papualthia Diels ......................................Annonaceae
Papuanthes Danser ...................................Loranthaceae

Papuapteris = Polystichum .......................................... Aspleniaceae (Dryopt.)

**Papuastelma** Bullock................................................. Asclepiadaceae

**Papuechites** Markgraf .............................................. Apocynaceae

Papuodendron see Hibiscus (Malvaceae) .......................... Bombacaceae

Papuzilla = Lepidium .................................................. Brassicaceae

**Parabaena** Miers .................................................... Menispermaceae

**Paragramma** (Blume) T. Moore................................... Polypodiaceae (Pleo.)

Paragulubia = Gulubia................................................ Arecaceae

**Parahebe** W. Oliver ................................................. Scrophulariaceae

Paralinospadix = Calyptrocalyx .................................... Arecaceae

Paralstonia = Alyxia .................................................. Apocynaceae

Paramapania = Mapania .............................................. Cyperaceae

**Paramigyna** Wight (Atalantia) .................................... Rutaceae

**Parartocarpus** Baillon .............................................. Moraceae

Parascopolia see Solanum............................................ Solanaceae

Parasorus Alderw..................................................... Davalliaceae (Davall.) .... no record

**Parasponia** Miq. .................................................... Ulmaceae

**Parastemon** A. DC................................................... Chrysobalanaceae

**Parathelypteris** (H. Itô) Ching (Thelypteris)..................... Thelypteridaceae

Paratrophis = Streblus ............................................... Moraceae

Paratropia = Schefflera .............................................. Araliaceae

**Parietaria** L. ......................................................... Urticaceae

**Parinari** Aublet ...................................................... Chrysobalanaceae

Parishia Hook. f. ...................................................... Anacardiaceae ............ no record

**Parkia** R. Br. ........................................................ Mimosaceae

**Parkinsonia** L. ....................................................... Caesalpiniaceae

**Parsonsia** R. Br. ..................................................... Apocynaceae

Pasania = Lithocarpus ................................................ Fagaceae

**Paspalidium** Stapf (Setaria)........................................ Poaceae

**Paspalum** L. .......................................................... Poaceae

**Passiflora** L. ......................................................... Passifloraceae

**Patersonia** R. Br. .................................................... Iridaceae

**Pavetta** L. ............................................................ Rubiaceae

**Payena** A. DC. ....................................................... Sapotaceae

Pedilanthus Necker ex Poit. (Euphorbia) .......................... Euphorbiaceae ............ no record

**Pedilochilus** Schltr. ................................................. Orchidaceae

Peekelia = Oxyrhynchus ............................................. Fabaceae

Peekeliodendron = Merrilliodendron ............................... Icacinaceae

Peekeliopanax = Gastonia............................................ Araliaceae

**Pellaea** Link......................................................... Adiantaceae (Adiant.)

**Pellionia** Gaudich. (Elatostema).................................. Urticaceae

Pelma = Bulbophyllum ............................................... Orchidaceae

**Peltophorum** (Vogel) Benth. ...................................... Caesalpiniaceae

**Pemphis** Forster & Forster f. ...................................... Lythraceae

Pennisetum Rich. ex Pers. ............................................ Poaceae

Pentaloba = Rinorea ................................................. Violaceae

Pentapanax Seemann ................................................. Araliaceae .................. no record

Pentapetes L. ........................................................ Sterculiaceae

Pentaphalangium Warb. (Garcinia).............................. Clusiaceae

Pentaphragma Wallich ex G. Don f. ............................ Pentaphragmataceae

Pentas Benth. ........................................................ Rubiaceae

Pentaspadon Hook. f. ............................................... Anacardiaceae

Pentastira = Dichapetalum ........................................ Dichapetalaceae

Pentatropis Wight & Arn. .......................................... Asclepiadaceae

Peperomia Ruíz & Pavón .......................................... Piperaceae

Peracarpa Hook. f. & Thomson .................................. Campanulaceae

Peranema D. Don ................................................... Aspleniaceae (Dryopt.) ... no record

Pericampylus Miers ................................................. Menispermaceae

Pericopsis Thwaites ................................................. Fabaceae

Peripterygium = Cardiopteris ..................................... Cardiopteridaceae

Peristrophe Nees..................................................... Acanthaceae

Peristylus Blume (Herminium) .................................... Orchidaceae

Perotis Aiton ......................................................... Poaceae

Perrottetia Kunth ................................................... Celastraceae

Persea Miller ......................................................... Lauraceae

Pertusadina Ridsd. (Adina)........................................ Rubiaceae

Petalolophus Schumann ............................................ Annonaceae

Petalostigma F. Muell. .............................................. Euphorbiaceae

Petraeovitex Oliver................................................... Verbenaceae

Petroselinum Hill.................................................... Apiaceae

Petunga = Hypobathrum ........................................... Rubiaceae

Phacellaria Benth. ................................................... Santalaceae

Phacellothrix F. Muell. ............................................. Asteraceae

Phacelophrynium Schumann ...................................... Marantaceae

Phaeanthus Hook. f. & Thomson ................................ Annonaceae

Phaeomeria = Nicolaia ............................................. Zingiberaceae

Phaius Lour........................................................... Orchidaceae

Phalaenopsis Blume ................................................ Orchidaceae

Phalaris see Arthraxon .............................................. Poaceae

Phaleria Jack ......................................................... Thymelaeaceae

Phanera Lour. ........................................................ Caesalpiniaceae

Phanerosorus Copel.................................................. Matoniaceae

Pharbitis = Ipomoea................................................. Convolvulaceae

Phaseolus L........................................................... Fabaceae

Phegopteris Fée ...................................................... Thelypteridaceae

Pheidochloa S.T. Blake.............................................. Poaceae

Phelline Labill. ...................................................... Aquifoliaceae ............. no record

Phillyrea see Chionanthus ............................................ Oleaceae

**Philodendron** Schott .................................................. Araceae

**Philydrum** Banks ex Gaertner........................................ Philydraceae

**Phlogacanthus** Nees ................................................. Acanthaceae

**Phoebe** Nees ........................................................ Lauraceae

Phoenicosperma = Sloanea .............................................. Elaeocarpaceae

**Pholidota** Lindley ex Hook. ......................................... Orchidaceae

Photinopteris see Merinthosorus ....................................... Polypodiaceae (Dryn.)

**Phragmites** Adans. ................................................... Poaceae

**Phreatia** Lindley .................................................... Orchidaceae

Phrygilanthus = Notanthera ............................................ Loranthaceae

**Phrynium** Willd. .................................................... Marantaceae

**Phyla** Lour. (Lippia) ............................................... Verbenaceae

**Phylacium** Bennett.................................................. Fabaceae

**Phyllanthera** Blume ................................................ Asclepiadaceae

**Phyllanthus** L. ..................................................... Euphorbiaceae

Phyllapophysis Mansf. = Catanthera .................................... Melastomataceae

Phyllitis = Asplenium................................................. Aspleniaceae (Asplen.)

Phyllocharis = Ruthiella............................................... Campanulaceae

**Phyllocladus** Rich. ex Mirbel ...................................... Phyllocladaceae

Phyllodium = Desmodium .............................................. Fabaceae

Phyllorchis = Bulbophyllum........................................... Orchidaceae

Phymatodes = Microsorum ............................................ Polypodiaceae (Micro.)

**Physalis** L.......................................................... Solanaceae

**Physokentia** Becc. .................................................. Arecaceae

**Physostelma** Wight ................................................. Asclepiadaceae

Physurus = Erythrodes ................................................ Orchidaceae

**Phytocrene** Wallich ................................................. Icacinaceae

**Phytolacca** L. ...................................................... Phytolaccaceae

**Pichonia** Pierre ..................................................... Sapotaceae

**Picrasma** Blume..................................................... Simaroubaceae

**Pigafetta** (Blume) Becc. ............................................ Arecaceae

Pigafetta Adans. = Erianthemum ...................................... Acanthaceae

**Pilea** Lindley ...................................................... Urticaceae

**Pilophyllum** Schltr. (Chrysoglossum) ............................... Orchidaceae

Pimelandra = Ardisia ................................................. Myrsinaceae

**Pimelea** Banks & Sol. ex Gaertner .................................. Thymelaeaceae

**Pimelodendron** Hassk. .............................................. Euphorbiaceae

**Pinanga** Blume ..................................................... Arecaceae

**Pinus** L. ........................................................... Pinaceae

**Piora** J. Koster ..................................................... Asteraceae

**Piper** L. ........................................................... Piperaceae

Piptadenia = Schleinitzia ............................................. Mimosaceae

**Piptocalyx** Oliver ex Benth. (Trimenia) ............................. Trimeniaceae

Piptospatha N.E. Br. ............................................... Araceae ...................... no record
**Pipturus** Wedd. ..................................................... Urticaceae
**Pisonia** L. ............................................................... Nyctaginaceae
**Pistia** L. ................................................................. Araceae
**Pisum** L. ................................................................. Fabaceae
**Pithecellobium** C. Martius ....................................... Mimosaceae
**Pittosporum** Banks ex Gaertner ................................... Pittosporaceae
**Pityrogramma** Link ................................................. Adiantaceae (Adiant.)
Plagiobothrys see Trigonotis ....................................... Boraginaceae
**Plagiogyria** (Kunze) Mett. ....................................... Plagiogyriaceae
Planchonella see Pichonia, Pouteria ............................. Sapotaceae
**Planchonia** Blume .................................................... Barringtoniaceae
**Plantago** L. ............................................................ Plantaginaceae
**Platanthera** Rich. (Habenaria) ...................................... Orchidaceae
**Platea** Blume ........................................................... Icacinaceae
**Platycerium** Desv. .................................................... Polypodiaceae (Platy.)
Platyclinis see Dendrochilum ....................................... Orchidaceae
**Platycoryne** Riechb. f. ............................................... Orchidaceae
**Platylepis** A. Rich. ................................................... Orchidaceae
**Platymitra** Boerl. ..................................................... Annonaceae
Platytaenia Kuhn (Taenitis) ......................................... Adiantaceae (Adiant.) ..... no record
**Platyzoma** R. Br. ..................................................... Platyzomataceae ........... no record
**Plectranthus** L'Hérit. ................................................ Lamiaceae
Plectronia see Canthium ............................................. Rubiaceae
**Pleiogynium** Engl. .................................................... Anacardiaceae
**Pleione** D. Don (Coelogyne) ....................................... Orchidaceae
**Pleocnemia** C. Presl .................................................. Aspleniaceae (Tect.)
Pleomele = Dracaena ................................................. Dracaenaceae
Pleopeltis see Crypsinus, Microsorum ........................... Polypodiaceae (Micro.)
Plerandra = Schefflera ................................................ Araliaceae
**Plesioneuron** (Holttum) Holttum (Thelypteris) .................. Thelypteridaceae
Pleurogramme = Monogramma ..................................... Adiantaceae (Vitt.)
**Pleuromanes** C. Presl (Trichomanes) ............................. Hymenophyllaceae
**Pleurostylia** Wight & Arn. .......................................... Celastraceae
*Plocoglottis* Blume ..................................................... Orchidaceae
**Ploiarium** Korth. ...................................................... Theaceae
**Pluchea** Cass. .......................................................... Asteraceae
**Plumbago** L. ........................................................... Plumbaginaceae
**Plumeria** L. ............................................................ Apocynaceae
**Pneumatopteris** Nakai ............................................... Thelypteridaceae
**Poa** L. ................................................................... Poaceae
**Poaephyllum** Ridley .................................................. Orchidaceae
**Poarium** Desv. (Stemodia) .......................................... Scrophulariaceae

Pocillaria = Rhyticaryum ............................................. Icacinaceae

**Podocarpus** L'Hérit. ex Pers. ..................................... Podocarpaceae

**Podochilus** Blume..................................................... Orchidaceae

**Podosorus** Holttum ................................................... Polypodiaceae (Micro.)

**Podostemum** Michaux ............................................... Podostemaceae

**Pogonanthera** Blume ................................................ Melastomataceae

**Pogonatherum** Pal..................................................... Poaceae

**Pogonia** Juss. (Nervilia) ........................................... Orchidaceae

Pogonolobus = Coelospermum ..................................... Rubiaceae

**Pogostemon** Desf. .................................................... Lamiaceae

**Poikilogyne** Baker f. ................................................ Melastomataceae

**Poikilospermum** Zipp. ex Miq. ................................... Cecropiaceae

Poinciana = Caesalpinia ............................................. Caesalpiniaceae

Polanisia = Cleome ................................................... Capparaceae

**Pollia** Thunb. ......................................................... Commelinaceae

Pollinia see Eulalia, Microstegium................................ Poaceae

**Polyalthia** Blume...................................................... Annonaceae

**Polyaulax** Backer ..................................................... Annonaceae

Polybotrya Humb. & Bonpl. ex Willd. .......................... Aspleniaceae (Dryopt.) ... no record

**Polycarpaea** Lam....................................................... Caryophyllaceae

**Polygala** L............................................................... Polygalaceae

**Polygonum** L. .......................................................... Polygonaceae

**Polyosma** Blume ...................................................... Grossulariaceae

Polyphragmon = Timonius .......................................... Rubiaceae

Polypodiopteris C. Reed ............................................. Polypodiaceae (Micro.)... no record

**Polypodium** L. ......................................................... Polypodiaceae (Poly.)

**Polyporandra** Becc. .................................................. Icacinaceae

**Polyscias** Forster & Forster f. .................................... Araliaceae

Polystichopsis (J. Sm.) Holttum.................................... Aspleniaceae (Dryopt.) ... no record

**Polystichum** Roth ..................................................... Aspleniaceae (Dryopt.)

**Polytoca** R. Br. ........................................................ Poaceae

Polytrema C.B. Clarke ................................................ Acanthaceae................ no record

Polytrias = Eulalia ..................................................... Poaceae

**Pomatocalpa** Breda, Kuhl & Hasselt ............................ Orchidaceae

**Pometia** Forster & Forster f. ....................................... Sapindaceae

**Poncirus** Raf. .......................................................... Rutaceae

**Pongamia** Vent......................................................... Fabaceae

Pontederia see Monochoria ......................................... Pontederiaceae

**Popowia** Endl. ......................................................... Annonaceae

**Porana** Burm. f. ....................................................... Convolvulaceae

Porotheca = Chlaenandra ........................................... Menispermaceae

**Porphyrodesme** Schltr................................................ Orchidaceae

Portula = Lythrum ..................................................... Lythraceae

**Portulaca** L. ............................................................ Portulacaceae

Potamogeton L. ........................................................... Potamogetonaceae
Potentilla L. ............................................................ Rosaceae
Pothomorphe = Lepianthes .......................................... Piperaceae
Pothos L. .............................................................. Araceae
Pouteria Aublet ...................................................... Sapotaceae
Pouzolzia Gaudich. .................................................. Urticaceae
Prainea King ex Hook. f. .......................................... Moraceae
Pratia = Lobelia...................................................... Lobeliaceae
Premna L............................................................... Verbenaceae
Primula L. ............................................................. Primulaceae
Pristiglottis Cretz. & J.J. Sm...................................... Orchidaceae
Pritchardia Seemann & H. Wendl. .............................. Arecaceae
Procris Comm. ex Juss. ............................................ Urticaceae
Proiphys Herbert...................................................... Amaryllidaceae
Pronephrium C. Presl ............................................... Thelypteridaceae
Prosaptia C. Presl ................................................... Grammitidaceae
Prosopis L............................................................... Mimosaceae
Protium Burm. f. ..................................................... Burseraceae
Prumnopitys = Podocarpus ........................................ Podocarpaceae
Prunus L................................................................ Rosaceae
Psacadocalymma = Stethoma ...................................... Acanthaceae
Pseudechinolaena Stapf ............................................. Poaceae
Pseudelephantopus Rohr (Elephantopus) ......................... Asteraceae
Pseuderanthemum Radlk. ........................................... Acanthaceae
Pseuderia Schltr....................................................... Orchidaceae
Pseudobotrys Moser .................................................. Icacinaceae
Pseudocarapa Hemsley (Dysolxylum) ............................ Meliaceae
Pseudochrosia = Ochrosia .......................................... Apocynaceae
Pseudocryptocarya = Cryptocarya ................................ Lauraceae
Pseudocyclosorus Ching.............................................. Thelypteridaceae
Pseudocynometra = Maniltoa ...................................... Caesalpiniaceae
Pseudoeugenia = Syzygium.......................................... Myrtaceae
Pseudoliparis see Malaxis ........................................... Orchidaceae
Pseudomacodes = Macodes ......................................... Orchidaceae
Pseudomorus = Streblus.............................................. Moraceae
Pseudophegopteris Ching ........................................... Thelypteridaceae
Pseudopinanga = Pinanga ........................................... Arecaceae
Pseudopipturus = Nothocnide....................................... Urticaceae
Pseudopogonatherum = Eulalia ................................... Poaceae
Pseudoraphis Griffith ................................................ Poaceae
Pseudosmelia Sleumer ............................................... Flacourtiaceae
Pseudotrophis = Streblus............................................ Moraceae
Pseudowillughbeia = Melodinus.................................... Apocynaceae

Pseuduvaria Miq.......................................................Annonaceae

Psidium L. ...........................................................Myrtaceae

Psilotrichum Blume ...............................................Amaranthaceae

Psilotum Sw...........................................................Psilotaceae

Psomiocarpa C. Presl .........................................Aspleniaceae (Tect.) ......no record

Psophocarpus Necker ex DC.....................................Fabaceae

Psoralea L.............................................................Fabaceae

Psychanthus = Alpinia............................................Zingiberaceae

Psychotria L. ........................................................Rubiaceae

Pteridium Gled. ex Scop. .........................................Dennstaedtiaceae (Denn.)

Pteridrys C. Chr. & Ching ......................................Aspleniaceae (Tect.)

Pteris L................................................................Adiantaceae (Pter.)

Pternandra Jack ....................................................Melastomataceae

Pterocarpus Jacq....................................................Fabaceae

Pterocaulon Elliott ................................................Asteraceae

Pteroceras Hasselt ex Hassk. (Sarcochilus).....................Orchidaceae

Pterocymbium R. Br................................................Sterculiaceae

Pteropsis = Pyrrosia ..............................................Polypodiaceae (Platy.)

Pterospermum Schreber............................................Sterculiaceae

Pterostelma = Hoya ...............................................Asclepiadaceae

Pterostylis R. Br. ..................................................Orchidaceae

Pterygota Schott & Endl. .........................................Sterculiaceae

Ptilotus R. Br. ......................................................Amaranthaceae

Ptychandra = Heterospathe .......................................Arecaceae

Ptychococcus Becc. .................................................Arecaceae

Ptychopyxis Miq.....................................................Euphorbiaceae

Ptychoraphis see Rhopaloblaste ..................................Arecaceae

Ptychosperma Labill. ..............................................Arecaceae

Ptyssiglottis T. Anderson .........................................Acanthaceae

Pueraria DC. ........................................................Fabaceae

Pullea Schltr. ........................................................Cunoniaceae

Punica L...............................................................Punicaceae

Pupalia Juss. ........................................................Amaranthaceae

Putranjiva = Drypetes .............................................Euphorbiaceae

Pycnarrhena Miers ex Hook. f. & Thomson.....................Menispermaceae

Pycnoloma C. Chr. ................................................Polypodiaceae (Micro.)...no record

Pycnospora R. Br. ex Wight & Arn. .............................Fabaceae

Pycreus Pal. (Cyperus) ............................................Cyperaceae

Pygeum = Prunus ..................................................Rosaceae

Pygmaeopremna = Premna ........................................Verbenaceae

Pyrostegia C. Presl .................................................Bignoniaceae

Pyrrhanthus = Lumnitzera .........................................Combretaceae

Pyrrosia Mirbel .....................................................Polypodiaceae (Platy.)

Pyrsonota = Sericolea ..............................................Elaeocarpaceae

Quamoclit = Ipomoea.................................................Convolvulaceae
**Quassia** L. .............................................................Simaroubaceae
Quercifilix = Tectaria................................................Aspleniaceae (Tect.)
Quercus see Lithocarpus .........................................Fagaceae
**Quintinia** A. DC. ...................................................Saxifragaceae
**Quisqualis** L. ........................................................Combretaceae
**Racemobambos** Holttum ......................................Poaceae
Radermachera Zoll. & Moritzi .....................................Bignoniaceae................no record
**Randia** L.................................................................Rubiaceae
**Rangaeris** (Schltr.) Summerh. ...............................Orchidaceae
**Ranunculus** L. ......................................................Ranunculaceae
**Raoulia** Hook. f. ex Raoul (Gnaphalium) ..............Asteraceae
**Rapanea** Aublet (Myrsine)....................................Myrsinaceae
**Raphanus** L...........................................................Brassicaceae
Raphidophora = Rhaphidophora ..............................Araceae
**Rauvolfia** L. .........................................................Apocynaceae
**Rauwenhoffia** R. Scheffer ....................................Annonaceae
**Ravenala** Adans. ..................................................Strelitziaceae
Rehderophoenix = Drymophloeus...............................Arecaceae
**Reinwardtiodendron** Koord. .................................Meliaceae
**Reissantia** Hallé....................................................Celastraceae
Rejoua = Tabernaemontana........................................Apocynaceae
**Remirea** Aublet (Cyperus) ...................................Cyperaceae
**Renanthera** Lour...................................................Orchidaceae
**Restio** Rottb. ........................................................Restionaceae
Retrophyllum see Decussocarpus ............................Podocarpaceae
Rhabdia = Rotula.......................................................Boraginaceae................no record
**Rhadinopus** S. Moore ..........................................Rubiaceae
**Rhamnus** L. .........................................................Rhamnaceae
**Rhamphicarpa** Benth. ..........................................Scrophulariaceae
**Rhamphogyne** S. Moore ......................................Asteraceae
**Rhaphidophora** Hassk. ........................................Araceae
**Rhaphidospora** Nees (Justicia)............................Acanthaceae
Rhaphis = Chrysopogon.............................................Poaceae
**Rheopteris** Alston.................................................Adiantaceae (Adiant.)
**Rhizomonanthes** Danser ......................................Loranthaceae
**Rhizophora** L. ......................................................Rhizophoraceae
**Rhodamnia** Jack ..................................................Myrtaceae
**Rhododendron** L. .................................................Ericaceae
**Rhodomyrtus** (DC.) Reichb. .................................Myrtaceae
Rhoeo = Tradescantia ...............................................Commelinaceae
**Rhopaloblaste** R. Scheffer ...................................Arecaceae
**Rhus** L. ...............................................................Anacardiaceae

**Rhynchelytrum** Nees ................................................ Poaceae
Rhynchocarpa = Kedrostis........................................... Cucurbitaceae
Rhynchodia Benth. ................................................... Apocynaceae................ no record
**Rhynchoglossum** Blume ............................................. Gesneriaceae
Rhynchophreatia = Phreatia ......................................... Orchidaceae
**Rhynchosia** Lour.................................................... Fabaceae
**Rhynchospora** Vahl ................................................. Cyperaceae
**Rhynchotoechum** Blume (Rhynchotechum) ...................... Gesneriaceae
**Rhysotoechia** Radlk. ............................................... Sapindaceae
**Rhyssopteris** Blume ex A. Juss. (Rhyssopterys) ................ Malpighiaceae
**Rhyticaryum** Becc.................................................. Icacinaceae
**Richardia** L........................................................ Rubiaceae
**Ricinus** L. ......................................................... Euphorbiaceae
**Ridleyella** Schltr. ................................................. Orchidaceae
**Riedelia** Oliver ................................................... Zingiberaceae
**Rinorea** Aublet .................................................... Violaceae
**Ripogonum** Forster & Forster f. (Rhipogonum) ................ Liliaceae
**Rivina** L. .......................................................... Phytolaccaceae
Robinsoniodendron = Maoutia...................................... Urticaceae
**Robiquetia** Gaudich. ............................................... Orchidaceae
**Rollinia** A. St-Hil.................................................. Annonaceae
**Romnalda** Harvey ................................................. Liliaceae
**Rorippa** Scop. ..................................................... Brassicaceae
**Rosa** L............................................................. Rosaceae
**Rotala** L. .......................................................... Lythraceae
**Rottboellia** L. f.................................................... Poaceae
Rotula Lour........................................................... Boraginaceae................ no record
**Rourea** Aublet ..................................................... Connaraceae
Roxburghia = Stemona ............................................. Stemonaceae
**Rubus** L. .......................................................... Rosaceae
**Ruellia** L. ......................................................... Acanthaceae
**Rumex** L. .......................................................... Polygonaceae
**Rumohra** Raddi .................................................... Davalliaceae (Davall.)
**Rungia** Nees ....................................................... Acanthaceae
**Ruppia** L. .......................................................... Ruppiaceae ................ no record
**Russelia** Jacq...................................................... Scrophulariaceae
**Ruthiella** Steenis .................................................. Campanulaceae
**Ryparosa** Blume ................................................... Flacourtiaceae
Ryssopterys = Rhyssopteris ........................................ Malpighiaceae
Sabia Colebr. ........................................................ Sabiaceae
**Saccharum** L. ...................................................... Poaceae
**Sacciolepis** Nash (Saccolepis) .................................... Poaceae
**Saccoglossum** Schltr. .............................................. Orchidaceae
**Saccolabiopsis** J.J. Sm. ........................................... Orchidaceae

Saccolabium Blume ................................................ Orchidaceae

Saccoloma Kaulf. .................................................. Dennstaedtiaceae (Denn.)

Saccopetalum = Miliusa .......................................... Annonaceae

Sagina L. ........................................................... Caryophyllaceae

Sagittaria L. ....................................................... Alismataceae

Salacia L. ........................................................... Celastraceae

Salacicratea Loes. (Salacia) .................................... Celastraceae

Salicornia L. ....................................................... Chenopodiaceae

Salmalia = Bombax ............................................... Bombacaceae

Salomonia Lour. ................................................... Polygalaceae

Salsola L. ........................................................... Chenopodiaceae

Salvia L. ............................................................ Lamiaceae

Salvinia Séguier ................................................... Salviniaceae

Samadera = Quassia .............................................. Simaroubaceae

Samanea (Benth.) Merr. .......................................... Mimosaceae

Sambucus L. ........................................................ Caprifoliaceae

Sanchezia Ruíz & Pavón .......................................... Acanthaceae

Sandoricum Cav. ................................................... Meliaceae

Sanicula L. ......................................................... Apiaceae

Sanseviera Thunb. ................................................. Sansevieraceae

Santaloides = Rourea ............................................. Connaraceae

Santalum L. ........................................................ Santalaceae

Santiria Blume ..................................................... Burseraceae

Sapindus L. ........................................................ Sapindaceae

Sapium see Excoecaria ............................................ Euphorbiaceae

Saprosma Blume .................................................... Rubiaceae

Saraca L. ........................................................... Caesalpiniaceae

Sararanga Hemsley ................................................ Pandanaceae

Sarcandra Gardner ................................................ Chloranthaceae

Sarcanthus = Cleisostoma ........................................ Orchidaceae

Sarcocephalus see Nauclea ....................................... Rubiaceae

Sarcochilus R. Br. ................................................. Orchidaceae

Sarcolobus R. Br. .................................................. Asclepiadaceae

Sarcopetalum F. Muell. ............................................ Menispermaceae

Sarcopodium see Dendrobium ..................................... Orchidaceae

Sarcopteryx Radlk. ................................................ Sapindaceae

Sarcosperma Hook. f. .............................................. Sapotaceae

Sarcotoechia ....................................................... Sapindaceae

Satureja L. ......................................................... Lamiaceae

Saurauia Willd. .................................................... Actinidiaceae

Sauropus Blume .................................................... Euphorbiaceae

Sayeria see Dendrobium ........................................... Orchidaceae

Scaevola L. ......................................................... Goodeniaceae

**Scaphiophora** Schltr. ............................................... Burmanniaceae

**Scaphium** Schott & Endl. ............................................ Sterculiaceae .............. no record

Schaueria see Hyptis (Lamiaceae)................................ Acanthaceae

**Schefferomitra** Diels ............................................... Annonaceae

**Schefflera** Forster & Forster f...................................... Araliaceae

Scheffleropsis = Schefflera ......................................... Araliaceae

**Schelhammera** R. Br................................................ Liliaceae

**Schismatoglottis** Zoll. & Moritzi................................... Araceae

Schistostigma = Cleistanthus ....................................... Euphorbiaceae

**Schizachyrium** Nees ............................................... Poaceae

**Schizaea** Sm........................................................ Schizaeaceae

Schizocaena J. Sm. ex Hook. ....................................... Cyatheaceae................ no record

Schizocasia see Alocasia ............................................. Araceae

Schizolepton = Taenitis ............................................. Adiantaceae (Adiant.)

Schizoloma = Lindsaea................................................ Dennstaedtiaceae (Lind.)

**Schizomeria** D. Don ................................................ Cunoniaceae

Schizoscyphus = Maniltoa............................................ Caesalpiniaceae

Schizosiphon = Maniltoa ............................................ Caesalpiniaceae

Schizospatha = Calamus.............................................. Arecaceae

**Schizostachyum** Nees............................................... Poaceae

Schizostege = Pteris ................................................. Adiantaceae (Pter.)

Schizotheca = Thalassia ............................................. Hydrocharitaceae

**Schleichera** Willd. ................................................. Sapindaceae

**Schleinitzia** Warb. ex Harms (Prosopis) ......................... Mimosaceae

Schmidelia = Allophylus ............................................. Sapindaceae

**Schoenorchis** Blume ............................................... Orchidaceae

**Schoenus** L......................................................... Cyperaceae

**Schumannianthus** Gagnepain ..................................... Marantaceae

**Schuurmansia** Blume .............................................. Ochnaceae

Schychowskya = Laportea............................................ Urticaceae

Sciadophyllum = Schefflera ......................................... Araliaceae

**Sciaphila** Blume ................................................... Triuridaceae

**Scindapsus** Schott.................................................. Araceae

Sciophila = Procris................................................... Urticaceae

**Scirpodendron** Zipp. ex Kurz..................................... Cyperaceae

**Scirpus** L. ......................................................... Cyperaceae

**Sclerandrium** Stapf & C. Hubb. (Germainia)..................... Poaceae

**Scleranthus** L...................................................... Caryophyllaceae

**Scleria** P. Bergius ................................................. Cyperaceae

**Scleroglossum** Alderw. (Grammitis) ............................. Grammitidaceae

Scleromelum = Scleropyrum ........................................ Santalaceae

**Scleropyrum** Arn................................................... Santalaceae

**Sclerotheca** A. DC. ................................................ Campanulaceae

Scolopendrium = Asplenium ........................................ Aspleniaceae (Asplen.)

Scolopia Schreber ..................................................... Flacourtiaceae
Scoparia L......................................................... Scrophulariaceae
Scortechinia = Neoscortechinia .................................. Euphorbiaceae
Scrobicularia Mansf. = Poikilogyne ............................. Melastomataceae
Scutellaria L....................................................... Lamiaceae
Scutinanthe Thwaites ............................................. Burseraceae
Scyphiphora Gaertner f. ........................................... Rubiaceae
Scyphularia Fée ..................................................... Davalliaceae (Davall.)
Sebastiania Sprengel .............................................. Euphorbiaceae
Secamone R. Br. ................................................... Asclepiadaceae
Sechium P. Browne................................................. Cucurbitaceae
Securidaca L. ...................................................... Polygalaceae
Securinega Comm. ex Juss. ....................................... Euphorbiaceae
Sehima Forssk....................................................... Poaceae
Selaginella Pal...................................................... Selaginellaceae
Selenodesmium (Prantl) Copel. (Trichomanes) ................. Hymenophyllaceae
Selliguea Bory ...................................................... Polypodiaceae (Micro.)
Semecarpus L. f. ................................................... Anacardiaceae
Senecio L. .......................................................... Asteraceae
Senna = Cassia...................................................... Caesalpiniaceae
Sepalosiphon Schltr. ............................................... Orchidaceae
Sepikea Schltr. ..................................................... Gesneriaceae
Serialbizzia = Albizia............................................... Mimosaceae
Serianthes Benth. .................................................. Mimosaceae
Sericolea Schltr..................................................... Elaeocarpaceae
Seringia see Keraudrenia............................................ Sterculiaceae
Sesamum L. ........................................................ Pedaliaceae
Sesbania Scop. ..................................................... Fabaceae
Sesuvium L.......................................................... Aizoaceae
Setaria Pal.......................................................... Poaceae
Shorea see Hopea................................................... Dipterocarpaceae
Shuteria Wight & Arn. ............................................. Fabaceae
Sida L. ............................................................. Malvaceae
Sideroxylon L. (Planchonella)...................................... Sapotaceae
Sigesbeckia L....................................................... Asteraceae
Sinoga S.T. Blake.................................................. Myrtaceae
Siphokentia Burret ................................................ Arecaceae .................. no record
Siphonandrium Schumann........................................... Rubiaceae
Siphonodon Griffith ............................................... Celastraceae
Sisyrinchium L. ................................................... Iridaceae
Skoliostigma = Spondias ........................................... Anacardiaceae
Sloanea L. ......................................................... Elaeocarpaceae
Smilax L............................................................ Smilacaceae

Smithia Aiton ............................................................. Fabaceae

Smythea Seemann ex A. Gray.................................... Rhamnaceae

Sogerianthe Danser .................................................. Loranthaceae

Solanum L. ............................................................... Solanaceae

Solenopsis C. Presl .................................................. Campanulaceae

Solenospermum = Lophopetalum .............................. Celastraceae

Solenostemon Thonn. (Plectranthus)............................ Lamiaceae

Solenostigma = Celtis ............................................. Ulmaceae

Sommieria Becc. ....................................................... Arecaceae

Sonchus L. ............................................................... Asteraceae

Sonerila Roxb........................................................... Melastomataceae

Sonneratia L. f.......................................................... Sonneratiaceae

Sophora L. ............................................................... Fabaceae

Sopubia Buch.-Ham. ex D. Don ................................ Scrophulariaceae

Sorghum Moench ...................................................... Poaceae

Soulamea Lam. ......................................................... Simaroubaceae

Soya = Glycine.......................................................... Fabaceae

Spanoghea = Alectryon.............................................. Sapindaceae

Sparganium L. .......................................................... Sparganiaceae

Spathidolepis Schltr.................................................. Asclepiadaceae

Spathiostemon Blume................................................ Euphorbiaceae

Spathiphyllum Schott ............................................... Araceae

Spathodea Pal. .......................................................... Bignoniaceae

Spathoglottis Blume.................................................. Orchidaceae

Spergula L. ............................................................... Caryophyllaceae

Spergularia (Pers.) Pers. & C. Presl .......................... Caryophyllaceae ........... no record

Sperlingia = Hoya..................................................... Asclepiadaceae

Spermacoce L. .......................................................... Rubiaceae

Sphaeranthus L. ....................................................... Asteraceae

Sphaerocaryum Nees ex Hook. f. .............................. Poaceae ..................... no record

Sphaerocionium C. Presl (Hymenophyllum)...................... Hymenophyllaceae

Sphaerophora = Morinda .......................................... Rubiaceae

Sphaeropteris = Cyathea ........................................... Cyatheaceae

Sphaerostephanos J. Sm. ........................................... Thelypteridaceae

Sphenoclea Gaertner ................................................. Sphenocleaceae

Sphenodesme Jack..................................................... Verbenaceae ............... no record

Sphenomeris Maxon (Odontosoria) ............................ Dennstaedtiaceae (Lind.)

Sphenostemon Baillon .............................................. Aquifoliaceae

Spiculaea Lindley...................................................... Orchidaceae................ no record

Spigelia L.................................................................. Loganiaceae

Spilanthes Jacq. ........................................................ Asteraceae

Spinifex L. ................................................................ Poaceae

Spiraea L. ................................................................. Rosaceae

Spiraeanthemum A. Gray .......................................... Cunoniaceae

Spiraeopsis = Caldcluvia ............................................. Cunoniaceae

**Spiranthes** Rich. ..................................................... Orchidaceae

**Spirodela** Schleiden ................................................ Lemnaceae

**Spondias** L. .......................................................... Anacardiaceae

Sponia = Trema....................................................... Ulmaceae

**Sporobolus** R. Br. .................................................. Poaceae

Stachycarpus = Prumnopitys ....................................... Podocarpaceae

**Stachys** L. ............................................................ Lamiaceae

**Stachytarpheta** Vahl ................................................ Verbenaceae

**Stackhousia** Sm. .................................................... Stackhousiaceae

**Staphylea** L. ......................................................... Staphyleaceae .............. no record

**Stauranthera** Benth. ............................................... Gesneriaceae

**Staurogyne** Wallich ................................................ Acanthaceae

Stauropsis = Trichoglottis .......................................... Orchidaceae

**Steganthera** Perkins ............................................... Monimiaceae

Stegosia = Rottboellia ............................................... Poaceae

**Stelechocarpus** (Blume) Hook. f. & Thomson .................... Annonaceae

**Stellaria** L........................................................... Caryophyllaceae

**Stemona** Lour. ...................................................... Stemonaceae

**Stemonurus** Blume ................................................. Icacinaceae

**Stenocarpus** R. Br. ................................................ Proteaceae

**Stenochlaena** J. Sm................................................ Blechnaceae

**Stenolepia** Alderw. ................................................ Aspleniaceae (Dryopt.)

**Stenosemia** C. Presl ............................................... Aspleniaceae (Tect.)

**Stenotaphrum** Trin. ................................................ Poaceae

**Stephania** Lour. .................................................... Menispermaceae

**Stephanotis** Thouars................................................ Asclepiadaceae

Stephegyne = Mitragyna ............................................. Rubiaceae

**Sterculia** L. ......................................................... Sterculiaceae

**Stereosandra** Blume ............................................... Orchidaceae

**Stethoma** Raf. ...................................................... Acanthaceae................ no record

**Sticherus** C. Presl ................................................. Gleicheniaceae

**Stictocardia** Hallier f. ............................................. Convolvulaceae

Stillingia see Excoecaria ............................................ Euphorbiaceae

Stipa see Spinifex .................................................... Poaceae

Stizolobium = Mucuna .............................................. Fabaceae

Stollaea = Caldcluvia ................................................ Cunoniaceae

**Streblus** Lour....................................................... Moraceae

**Strelitzia** Banks ex Dryander .................................... Strelitziaceae

**Streptomanes** Schumann ex Schltr. .............................. Asclepiadaceae

**Striga** Lour.......................................................... Scrophulariaceae

**Strobilanthes** Blume................................................ Acanthaceae

**Strobilopanax** R. Viguier ......................................... Araliaceae .................. no record

# I. List of the Genera in New Guinea and the Solomon Islands

Strombosia Blume......................................................Olacaceae....................no record

Strongylocaryum = Ptychosperma ................................Arecaceae

**Strongylodon** Vogel ...............................................Fabaceae

Strophanthus see Papuechites ......................................Apocynaceae

**Struchium** P. Browne...............................................Asteraceae

**Strychnos** L. ........................................................Loganiaceae

**Stylidium** Sw. ex Willd. ..........................................Stylidiaceae

Stylocoryna = Tarenna ...............................................Rubiaceae

**Stylosanthes** Sw....................................................Fabaceae

**Styphelia** Sm. ......................................................Epacridaceae

**Styrax** L. ...........................................................Styracaceae

**Suaeda** Forssk. ex Scop. ..........................................Chenopodiaceae...........no record

**Suregada** Roxb. ex Rottler ......................................Euphorbiaceae

**Suriana** L. ..........................................................Surianaceae

Susum = Hanguana.....................................................Hanguanaceae

**Swertia** L. .........................................................Gentianaceae

**Swietenia** Jacq. ...................................................Meliaceae

**Sycopsis** Oliver (Distyliopsis) .................................Hamamelidaceae

Symbegonia = Begonia ...............................................Begoniaceae

**Symplocos** Jacq. ..................................................Symplocaceae

**Syncarpia** Ten. (Metrosideros)...................................Myrtaceae

**Syndyophyllum** Schumann & Lauterb. ............................Euphorbiaceae

**Synedrella** Gaertner ..............................................Asteraceae

**Syngonium** Schott .................................................Araceae

**Syngramma** J. Sm...................................................Adiantaceae (Adiant.)

**Synima** Radlk. .....................................................Sapindaceae

Synostemon see Sauropus ...........................................Euphorbiaceae

**Synoum** A. Juss. ...................................................Meliaceae....................no record

Syntherisma = Digitaria .............................................Poaceae

**Syringodium** Kütz..................................................Cymodoceaceae ...........no record

**Syzygium** Gaertner ................................................Myrtaceae

**Tabebuia** Gomes ex DC............................................Bignoniaceae

**Tabernaemontana** L. ...............................................Apocynaceae

**Tacca** Forster & Forster f. .......................................Taccaceae

Tacsonia = Passiflora.................................................Passifloraceae

**Taeniophyllum** Blume .............................................Orchidaceae

**Taenitis** Willd. ex Schkuhr ......................................Adiantaceae (Adiant.)

**Tagetes** L. .........................................................Asteraceae

**Tainia** Blume ......................................................Orchidaceae

Talauma = Magnolia .................................................Magnoliaceae

**Talinum** Adans......................................................Portulacaceae

**Tamarindus** L. .....................................................Caesalpiniaceae

**Tapeinidium** (C. Presl) C. Chr. ..................................Dennstaedtiaceae (Lind.)

**Tapeinochilos** Miq. ...............................................Zingiberaceae

Tapeinoglossum = Bulbophyllum ................................... Orchidaceae
**Tapeinosperma** Hook. f. ............................................... Myrsinaceae
**Taraxacum** Wigg. ..................................................... Asteraceae
**Tarenna** Gaertner ...................................................... Rubiaceae
Tarrietia = Heritiera ..................................................... Sterculiaceae
Tasmannia = Drimys ..................................................... Winteraceae
**Tecoma** Juss............................................................. Bignoniaceae
**Tecomanthe** Baillon .................................................. Bignoniaceae
**Tectaria** Cav. .......................................................... Aspleniaceae (Tect.)
**Tecticornia** Hook. f. .................................................. Chenopodiaceae
**Tectona** L. f. ........................................................... Verbenaceae
**Teijsmanniodendron** Koord. ...................................... Verbenaceae
Teleianthera = Alternanthera ........................................ Amaranthaceae
Tenagocharis = Butomopsis ........................................... Limnocharitaceae
**Tephrosia** Pers. ....................................................... Fabaceae
**Teramnus** P. Browne ................................................ Fabaceae
**Teratophyllum** Mett.................................................. Aspleniaceae (Lom.)
**Terminalia** L. ......................................................... Combretaceae
Terminthodia = Tetractomia........................................... Rutaceae
**Ternstroemia** Mutis ex L. f. ...................................... Theaceae
**Tetracera** L. ........................................................... Dilleniaceae
**Tetractomia** Hook. f. (Melicope)................................ Rutaceae
Tetradenia = Neolitsea ................................................. Lauraceae
**Tetradyas** Danser ..................................................... Loranthaceae
Tetraglochidion = Glochidion ........................................ Euphorbiaceae
**Tetragonia** L. ......................................................... Aizoaceae
**Tetrameles** R. Br. .................................................... Datiscaceae
**Tetramolopium** Nees ................................................ Asteraceae
Tetranthera = Litsea .................................................... Lauraceae
**Tetraphyllum** Griffith ex C.B. Clarke........................... Gesneriaceae
Tetraplasandra = Gastonia ............................................ Araliaceae
Tetrastigma (Miq.) Planchon.......................................... Vitaceae
Tetrathalamus see Drimys .............................................. Winteraceae
**Teucrium** L. ........................................................... Lamiaceae
**Thalassia** Banks & Sol. ex C. Koenig .......................... Hydrocharitaceae
Thalia see Donax ......................................................... Marantaceae
**Thalictrum** L. ......................................................... Ranunculaceae
**Thaumastochloa** C. Hubb. ......................................... Poaceae
**Thayeria** Copel. ...................................................... Polypodiaceae (Dryn.)
**Thelasis** Blume ....................................................... Orchidaceae
**Thelymitra** Forster & Forster f. .................................. Orchidaceae
**Thelypteris** Schmidel ............................................... Thelypteridaceae
**Themeda** Forssk. ..................................................... Poaceae

# I. List of the Genera in New Guinea and the Solomon Islands

| | |
|---|---|
| **Theobroma** L. | Sterculiaceae |
| **Thespesia** Sol. ex Corr. Serr. | Malvaceae |
| **Thevetia** L. | Apocynaceae |
| **Thismia** Griffith | Burmanniaceae |
| Thoa = Gnetum | Gnetaceae |
| Thoracostachyum = Mapania | Cyperaceae |
| **Thrixspermum** Lour. | Orchidaceae |
| **Thuarea** Pers. (Thouarea) | Poaceae |
| **Thunbergia** Retz. | Acanthaceae |
| **Thylacophora** Ridley (Riedelia) | Zingiberaceae |
| **Thylacopteris** Kunze ex J. Sm. | Polypodiaceae (Poly.) |
| **Thysanolaena** Nees | Poaceae |
| **Thysanosoria** Gepp | Aspleniaceae (Lom.) |
| **Thysanotus** R. Br. | Liliaceae |
| Tiaridium = Heliotropium | Boraginaceae |
| Tieghemopanax = Polyscias | Araliaceae |
| **Timonius** DC. | Rubiaceae |
| **Tinomiscium** Miers ex Hook. f. | Menispermaceae |
| **Tinospora** Miers | Menispermaceae |
| **Tipuana** (Benth.) Benth. | Fabaceae |
| **Tithonia** Desf. ex Juss. | Asteraceae |
| **Tmesipteris** Bernh. | Psilotaceae |
| **Todea** Willd. ex Bernh. | Osmundaceae |
| **Toechima** Radlk. | Sapindaceae |
| **Toona** (Endl.) M. Roemer | Meliaceae |
| **Torenia** L. | Scrophulariaceae |
| **Torrenticola** Domin ex Steenis | Podostemaceae |
| Torulinium = Cyperus | Cyperaceae |
| **Tournefortia** L. | Boraginaceae |
| **Toxocarpus** Wight & Arn. | Asclepiadaceae |
| **Trachoma** Garay | Orchidaceae |
| Trachylobium Hayne (Hymenaea) | Caesalpiniaceae ............ no record |
| **Trachymene** Rudge | Apiaceae |
| **Tradescantia** L. | Commelinaceae |
| **Trema** Lour. | Ulmaceae |
| Tremanthera = Saurauia | Actinidiaceae |
| Tremanthera = Saurauia (Actinidiaceae) | Theaceae |
| **Trevesia** Vis. | Araliaceae ................ no record |
| **Triadodaphne** Kosterm. | Lauraceae |
| **Trianthema** L. | Aizoaceae |
| **Tribulus** L. | Zygophyllaceae |
| **Tricalysia** A. Rich. ex DC. | Rubiaceae |
| **Trichadenia** Thwaites | Flacourtiaceae |
| Trichilia P. Browne | Meliaceae .................. no record |

Trichodesma R. Br. ................................................. Boraginaceae
Trichoglottis Blume ................................................ Orchidaceae
Tricholaena see Rhynchelytrum ................................... Poaceae
Trichomanes L. ...................................................... Hymenophyllaceae
Trichosanthes L. ..................................................... Cucurbitaceae
Trichospermum Blume .............................................. Tiliaceae
Trichosporum = Aeschynanthus ................................. Gesneriaceae
Trichotosia see Eria ................................................ Orchidaceae
Tricoryne R. Br. ..................................................... Liliaceae
Tricostularia Nees ex Lehm. ...................................... Cyperaceae
Tridax L. .............................................................. Asteraceae
Trifolium L. ........................................................... Fabaceae
Triglochin L. .......................................................... Juncaginaceae
Trigonachras Radlk. ................................................ Sapindaceae
Trigonostemon Blume .............................................. Euphorbiaceae
Trigonotis Steven .................................................... Boraginaceae
Trilocularia = Balanops ........................................... Balanopaceae .............. no record
Trimenia Seemann ................................................... Trimeniaceae
Tripetalum Schumann = Garcinia ................................ Clusiaceae
Triphasia Lour. ...................................................... Rutaceae
Triphlebia = Diplora ............................................... Aspleniaceae (Asplen.)
Triplochiton Schumann ............................................. Sterculiaceae
Triplostegia Wallich ex DC. ...................................... Valerianaceae
Tripogon Roemer & Schultes ...................................... Poaceae
Tripsacum L. .......................................................... Poaceae
Triraphis R. Br. ..................................................... Poaceae
Trisetum Pers. ........................................................ Poaceae
Tristania R. Br. (Tristaniopsis) ................................. Myrtaceae
Tristegis = Melinis ................................................. Poaceae
Tristellateia Thouars ............................................... Malpighiaceae
Tristira Radlk. ....................................................... Sapindaceae
Tristiropsis Radlk. .................................................. Sapindaceae
Tritonia Ker-Gawler ................................................ Iridaceae
Triumfetta L. ......................................................... Tiliaceae
Trochocarpa R. Br. ................................................. Epacridaceae
Trogostolon Copel. ................................................. Davalliaceae (Davall.) .... no record
Tropaeolum L. ....................................................... Tropaeolaceae ............. no record
Trophis P. Browne .................................................. Moraceae
Tropidia Lindley ..................................................... Orchidaceae
Turnera L. ............................................................. Turneraceae
Turpinia Vent. ........................................................ Staphyleaceae
Turraea L. ............................................................. Meliaceae
Tylecarpus = Medusanthera ...................................... Icacinaceae

**Tylophora** R. Br. (Riedelia) .......................................... Asclepiadaceae
**Tylosema** (Schweinf.) Torre & Hillcoat ........................... Caesalpiniaceae
**Typha** L. ................................................................ Typhaceae
**Typhonium** Schott ................................................... Araceae
**Uncaria** Schreber ..................................................... Rubiaceae
**Uncinia** Pers. ......................................................... Cyperaceae
Ungula = Amyema ...................................................... Loranthaceae
Unona = Xylopia ....................................................... Annonaceae
Uragoga = Cephaelis .................................................. Rubiaceae
**Urandra** Thwaites (Stemonurus) ................................... Icacinaceae
**Uraria** Desv. .......................................................... Fabaceae
**Urceola** Roxb. ........................................................ Apocynaceae
**Urena** L. ............................................................... Malvaceae
Urochloa see Alloteropsis, Brachiaria, Panicum ................. Poaceae
**Uromyrtus** Burret ..................................................... Myrtaceae
**Urophyllum** Jack ex Wallich ....................................... Rubiaceae
Urostachys = Lycopodium ............................................ Lycopodiaceae
**Urtica** L. ............................................................... Urticaceae
Uruparia = Uncaria .................................................... Rubiaceae
**Utricularia** L. ......................................................... Lentibulariaceae
**Uvaria** L. .............................................................. Annonaceae
**Vaccinium** L. .......................................................... Ericaceae
**Vaginularia** Fée ...................................................... Adiantaceae (Vitt.)
**Vallisneria** L. ......................................................... Hydrocharitaceae
**Vanda** Jones ex R. Br. ............................................... Orchidaceae
**Vandasina** Rauschert (Vandasia) ................................. Fabaceae
Vandellia = Lindernia ................................................. Scrophulariaceae
**Vandenboschia** Copel. (Trichomanes) ........................... Hymenophyllaceae
**Vandopsis** Pfitzer .................................................... Orchidaceae
Vangueria see Canthium .............................................. Rubiaceae
**Vanilla** Miller ......................................................... Orchidaceae
Vateria L. ............................................................... Dipterocarpaceae .......... no record
**Vatica** L. .............................................................. Dipterocarpaceae
**Vavaea** Benth. ........................................................ Meliaceae
**Veitchia** H. Wendl. ................................................... Arecaceae
**Velleia** Sm. ............................................................ Goodeniaceae
**Ventilago** Gaertner .................................................. Rhamnaceae
**Verbena** L. ............................................................ Verbenaceae
**Verbesina** L. .......................................................... Asteraceae
**Vernonia** Schreber ................................................... Asteraceae
**Veronica** L. ........................................................... Scrophulariaceae
**Versteegia** Valeton .................................................. Rubiaceae
**Vetiveria** Bory ........................................................ Poaceae
**Viburnum** L. .......................................................... Caprifoliaceae

Vicia L. ................................................................ Fabaceae

Vigna Savi ............................................................ Fabaceae

Villaresia = Citronella ............................................ Icacinaceae

Villaresia see Gonocaryum (Icacinaceae) ......................... Olacaceae

Villarsia Vent. ...................................................... Menyanthaceae

Villebrunea Gaudich. ex Wedd. ................................... Urticaceae

Vinca see Catharanthus.............................................. Apocynaceae

Vincetoxicum Wolf.................................................... Asclepiadaceae

Viola L. ............................................................... Violaceae

Viscum L. ............................................................ Viscaceae

Visiania = Ligustrum ............................................... Oleaceae

Vitex L. .............................................................. Verbenaceae

Viticipremna = Vitex ............................................... Verbenaceae

Vitis L. .............................................................. Vitaceae

Vittadinia A. Rich.................................................... Asteraceae

Vittaria Sm........................................................... Adiantaceae (Vitt.)

Voacanga Thouars .................................................... Apocynaceae

Vonroemeria = Octarrhena .......................................... Orchidaceae

Vrydagzynea Blume.................................................. Orchidaceae

Wahlenbergia Schrader ex Roth .................................... Campanulaceae

Walsura see Trichilia ................................................ Meliaceae

Waltheria L........................................................... Sterculiaceae

Weatherbya = Lemmaphyllum ....................................... Polypodiaceae (Pleo.)

Wedelia Jacq. (Wollastonia) ....................................... Asteraceae

Weinmannia L. ...................................................... Cunoniaceae

Wendlandia Bartling ex DC. ....................................... Rubiaceae

Wenzelia Merr........................................................ Rutaceae

Wetria Baillon........................................................ Euphorbiaceae

Whitmorea Sleumer................................................... Icacinaceae

Wibelia = Tapeinidium............................................... Dennstaedtiaceae (Lind.)

Wikstroemia Endl.................................................... Thymelaeaceae

Wilhelminia = Hibiscus ............................................ Malvaceae

Wilkiea F. Muell..................................................... Monimiaceae

Willughbeia Roxb. .................................................. Apocynaceae

Wittsteinia F. Muell. ............................................... Alseuosmiaceae ........... no record

Woikoia (Wokoia) = Pouteria....................................... Sapotaceae

Wolffia Horkel ex Schleiden........................................ Lemnaceae

Wollastonia DC. ex Decne. (Wedelia) ............................. Asteraceae ................. no record

Woodwardia Sm...................................................... Blechnaceae

Wormia = Dillenia .................................................. Dilleniaceae

Wrigthia R. Br. ..................................................... Apocynaceae

Xanthium L...../..................................................... Asteraceae ................. no record

Xanthochymus = Garcinia ........................................... Clusiaceae

# I. List of the Genera in New Guinea and the Solomon Islands

**Xanthomyrtus** Diels .................................................. Myrtaceae

**Xanthophyllum** Roxb. ............................................... Xanthophyllaceae

**Xanthophytum** Reinw. ex Blume ................................. Rubiaceae

**Xanthosoma** Schott................................................... Araceae

**Xanthostemon** F. Muell. ............................................ Myrtaceae

Xenodendron = Syzygium ............................................ Myrtaceae

Xenophya see Alocasia................................................. Araceae

Xeranthemum L.......................................................... Asteraceae ................. no record

**Xerocarpa** H.J. Lam = Teijsmanniodendron ..................... Verbenaceae

Xerotes = Lomandra ................................................... Xanthorrhoeaceae

**Ximenia** L. .............................................................. Olacaceae

**Xiphopteris** Kaulf. ................................................... Grammitidaceae

**Xylia** Benth. (Esclerona)............................................ Mimosaceae

**Xylocarpus** Koenig ................................................... Meliaceae

**Xylonymus** Kalkman ex Ding Hou ............................... Celastraceae

**Xylopia** L. .............................................................. Annonaceae

**Xylosma** Forster f. ................................................... Flacourtiaceae

**Xyris** L. ................................................................. Xyridaceae

**Youngia** Cass. ......................................................... Asteraceae

**Yucca** L. ................................................................ Agavaceae

**Zamioculcas** Schott .................................................. Araceae

**Zanonia** L. .............................................................. Cucurbitaceae

**Zantedeschia** Sprengel (Richardia) ............................... Araceae

**Zanthoxylum** L. (Xanthoxylum) .................................. Rutaceae

**Zea** L. ................................................................... Poaceae

Zebrina = Tradescantia................................................ Commelinaceae

**Zehneria** Endl. ........................................................ Cucurbitaceae

**Zephyranthes** Herbert ............................................... Amaryllidaceae

**Zeuxine** Lindley ...................................................... Orchidaceae

**Zingiber** Boehmer..................................................... Zingiberaceae

**Zinnia** L. ............................................................... Asteraceae

**Ziziphus** Miller (Zizyphus) ........................................ Rhamnaceae

Zoelleria = Trigonotis.................................................. Boraginaceae

**Zornia** J. Gmelin...................................................... Fabaceae

**Zoysia** Willd. .......................................................... Poaceae

**Zygogynum** Baillon ................................................... Winteraceae

## ADIANTACEAE (Adiantoidea)                12(43) / 43(742)

| | | |
|---|---|---|
| 086 | **Adiantum** L. | 6 / 11 / 200 |
| 062 | **Anogramma** Link | 0 / 1 / 7 |
| 085 | Cerosora (Baker) Domin | no record |
| 047 | **Cheilanthes** Sw. | 1 / 5 / 180 |
| 082 | **Coniogramme** Fée | 1 / 2 / 20 |
| | Craspedodictyum = Syngramma | |
| 044 | Cryptogramma R. Br. | no record |
| 058 | **Doryopteris** J. Sm. | 0 / 2 / 35 |
| | Gymnogramma = Gymnopteris | |
| 067 | **Gymnopteris** Bernh. (Hemionotis) | 2 / 2 / 5 |
| 066 | Hemionitis L. | no record |
| | Neurogramma = Gymnopteris | |
| 053 | Notholaena R. Br. (Cheilanthes) | no record |
| 043 | **Onychium** Kaulf. | 0 / 2 / 6 |
| 057 | **Pellaea** Link | 0 / 2 / 80 |
| 063 | **Pityrogramma** Link | 0 / 1 / 14 |
| 080 | Platytaenia Kuhn (Taenitis) | no record |
| 084 | **Rheopteris** Alston | 0 / 1 / 1    endemic in New Guinea |
| | Schizolepton = Taenitis | |
| 075 | **Syngramma** J. Sm. | 6 / 6 / 20 |
| 079 | **Taenitis** Willd. ex Schkuhr | 5 / 8 / 15 |

## ADIANTACEAE (Pteridoideae)                2(42) / 4(256)

| | | |
|---|---|---|
| 100 | **Acrostichum** L. | 4 / 4 / 4 |
| | Chrysodium see Acrostichum | |
| | Hemipteris = Pteris | |
| 096 | **Pteris** L. | 12 / 38 / 250 |
| | Schizostege = Pteris | |

## ADIANTACEAE (Vittarioideae)                4(37) / 9(102)

| | | |
|---|---|---|
| 087 | **Antrophyum** Kaulf. | 6 / 12 / 30 |
| 094 | **Monogramma** Comm. ex Schkuhr | 1 / 1 / 2 |
| | Pleurogramme = Monogramma | |
| 095 | **Vaginularia** Fée | 1 / 5 / 6 |
| 093 | **Vittaria** Sm. | 6 / 19 / 50 |

Angiopteridaceae see Marattiaceae

## ASPLENIACEAE (Asplenioideae)                3(49) / 11(670)

| | | |
|---|---|---|
| 288 | **Asplenium** L. | 32 / 44 / 650 |
| | Darea = Asplenium | |
| 298 | **Diplora** Baker (Asplenium) | 5 / 4 / 4 |
| 294 | **Loxoscaphe** T. Moore (Asplenium) | 0 / 1 / 4 |
| | Phyllitis = Asplenium | |

Scolopendrium = Asplenium
Triphlebia = Diplora

## ASPLENIACEAE (Athyrioideae)                      7(33) / 20(660)

| 313 | Anisocampium C. Presl. | no record |
|---|---|---|
| | Anisogonium = Diplazium | |
| 304 | **Athyrium** Roth | 2 / 12 / 180 |
| 309 | **Callipteris** Bory (Diplazium) | 0 / 2 / 3 |
| 305 | Cornopteris Nakai | no record |
| | Currania = Gymnocarpium | |
| 316 | **Cystopteris** Bernh. | 0 / 1 / 18 |
| 304 | Deparia Hook. & Grev. (Athyrium) | no record |
| 311 | **Diplaziopsis** C. Chr. (Diplazium) | 0 / 1 / 4 |
| 307 | **Diplazium** Sw. | 9 / 15 / 400 |
| | Dryoathyrium = Deparia | |
| 315 | **Gymnocarpium** Newman | 0 / 1 / 6 |
| 325 | **Hypodematium** Kunze | 0 / 1 / 3 |
| | Lunathyrium = Diplazium | |

## ASPLENIACEAE (Dryopteridoideae)                   6(45) / 16(380)

| 372 | **Acrophorus** C. Presl. | 1 / 1 / 2 |
|---|---|---|
| 364 | **Arachniodes** Blume | 0 / 1 / 20 |
| 355 | **Diacalpe** Blume | 0 / 1 / 1 |
| 369 | **Dryopteris** Adans. | 25 / 25 / 150 |
| | Nephrodium = Dryopteris | |
| | Papuapteris = Polystichum | |
| 354 | Peranema D. Don | no record |
| 367 | Polybotrya Humb. & Bonpl. ex Willd. | no record |
| 365 | Polystichopsis (J. Sm.) Holttum | no record |
| 356 | **Polystichum** Roth | 1 / 15 / 135 |
| 373 | **Stenolepia** Alderw. | 0 / 1 / 1 |

## ASPLENIACEAE (Elaphoglossoideae)                  1(38) / 1(400)

| 381 | Elaphoglossum Schott ex J. Sm. | 3 / 38 / 400 |
|---|---|---|

## ASPLENIACEAE (Lomariopsoideae)                    6(28) / 6(130)

| 377 | **Arthrobothrya** J. Sm. | 0 / 1 / 3 | |
|---|---|---|---|
| 374 | **Bolbitis** Schott | 7 / 7 / 47 | |
| | Campium = Bolbitis | | |
| | Egenolfia = Bolbitis | | |
| | Heteroneuron = Bolbitis | | |
| 379 | **Lomagramma** J. Sm. | 1 / 8 / 18 | |
| 380 | **Lomariopsis** Fée | 1 / 6 / 45 | |
| 378 | **Teratophyllum** Mett. | 2 / 4 / 12 | |
| 376 | **Thysanosoria** Gepp | 0 / 2 / 2 | endemic in New Guinea |

## ASPLENIACEAE (Tectarioideae)      10(69) / 24(460)

| | | |
|---|---|---|
| | Arcypteris = Pleocnemia | |
| | Aspidium = Tectaria | |
| 328 | **Ctenitis** (C. Chr.) Tard. & C. Chr. | 0 / 16 / 150 |
| 352 | **Cyclopeltis** J. Sm. | 1 / 2 / 6 |
| | Dictyopteris = Pleocnemia | |
| 353 | **Didymochlaena** Desv. | 1 / 2 / 2 |
| 336 | **Dryopolystichum** Copel. | 1 / 1 / 1     endemic in New Guinea |
| | Hemigramma = Tectaria | |
| 351 | **Heterogonium** C. Presl. | 0 / 3 / 20 |
| 330 | **Lastreopsis** Ching | 0 / 3 / 25 |
| 338 | Luerssenia Kuhn ex Luerssen (Tectaria) | no record |
| 333 | **Pleocnemia** C. Presl | 4 / 7 / 19 |
| 331 | Psomiocarpa C. Presl | no record |
| 335 | **Pteridrys** C. Chr. & Ching | 0 / 1 / 8 |
| | Quercifilix = Tectaria | |
| 350 | **Stenosemia** C. Presl | 0 / 1 / 2 |
| 337 | **Tectaria** Cav. | 13 / 34 / 200 |

## AZOLLACEAE      1(1) / 1(6)

| | | |
|---|---|---|
| 414 | **Azolla** Lam. | 0 / 1 / 6 |

## BLECHNACEAE      4(23) / 10(260)

| | | |
|---|---|---|
| 399 | **Blechnum** L. | 7 / 15 / 220 |
| 402 | Brainea J. Sm. | no record |
| 401 | **Doodia** R. Br. | 0 / 2 / 11 |
| | Lomaria = Blechnum | |
| 409 | **Stenochlaena** J. Sm. | 5 / 5 / 5 |
| 404 | **Woodwardia** Sm. | 0 / 1 / 12 |

## CHEIROPLEURIACEAE      1(1) / 1(1)

| | | |
|---|---|---|
| 151 | **Cheiropleuria** C. Presl. | 0 / 1 / 1 |

Christenseniaceae see Marattiaceae

Cryptogrammaceae see Adiantaceae (Adiantoideae)

## CYATHEACEAE      3(98) / 4(650)

| | | |
|---|---|---|
| | Alsophila = Cyathea | |
| | Balantium = Dicksonia | |
| 228 | **Cyathea** Sm. | 15 / 82 / 600 |
| 234 | **Cystodium** J. Sm. | 1 / 1 / 1 |
| 233 | **Dicksonia** L'Hérit. | 3 / 15 / 25 |
| | Gymnosphaera = Cyathea | |
| | Hemitelia = Cyathea | |
| 232 | Schizocaena J. Sm. ex Hook. | no record |
| | Sphaeropteris = Cyathea | |

## DAVALLIACEAE (Davallioideae)     6(56) / 11(130)

| | | |
|---|---|---|
| 391 | Araiostegia Copel. | no record |
| 388 | **Davallia** Sm. | 5 / 26 / 35 |
| 389 | **Davallodes** (Copel.) Copel. | 0 / 2 / 9 |
| 384 | **Humata** Cav. | 9 / 25 / 50 |
| 392 | **Leucostegia** C. Presl | 0 / 1 / 2 |
| 387 | Parasorus Alderw. | no record |
| 394 | **Rumohra** Raddi | 0 / 1 / 6 | cult. orn. |
| 386 | **Scyphularia** Fée | 1 / 1 / 8 |
| 385 | Trogostolon Copel. | no record |

## DAVALLIACEAE (Oleandriodeae)     3(41) / 4(91)

| | | |
|---|---|---|
| 396 | **Arthropteris** J. Sm. ex Hook. f. | 0 / 10 / 20 |
| 398 | **Nephrolepis** Schott | 9 / 20 / 30 |
| 395 | **Oleandra** Cav. | 5 / 11 / 40 |

## DENNSTAEDTIACEAE (Dennstaedtioideae)     10(54) / 9(210)

| | | |
|---|---|---|
| 238 | **Dennstaedtia** Bernh. | 3 / 19 / 70 |
| 245 | **Histiopteris** (J. Agardh) J. Sm. | 3 / 6 / 7 |
| 242 | **Hypolepis** Bernh. | 2 / 8 / 45 |
| | Ithyocaulon = Orthiopteris | |
| | Lepidocaulon see Histiopteris (Dennstaedtiaceae) | |
| 241 | **Leptolepia** Prantl (Microlepia) | 0 / 2 / 2 |
| 239 | **Microlepia** C. Presl | 1 / 9 / 45 | cult. orn. |
| 240 | **Oenotrichia** Copel. (Microlepia) | 0 / 1 / 4 |
| 250 | **Orthiopteris** Copel. (Saccoloma) | 2 / 5 / 9 |
| 243 | **Paesia** J. St-Hil. | 0 / 2 / 12 |
| 244 | **Pteridium** Gled. ex Scop. | 1 / 1 / 1 |
| 249 | **Saccoloma** Kaulf. | 0 / 1 / 1 |

## DENNSTAEDTIACEAE (Lindsaeoideae)     4(78) / 6(195)

| | | |
|---|---|---|
| | Isoloma see Lindsaea | |
| 252 | **Lindsaea** Dryander ex Sm. | 22 / 57 / 150 |
| 257 | **Odontosoria** Fée | 0 / 5 / 12 |
| | Schizoloma = Lindsaea | |
| 256 | **Sphenomeris** Maxon (Odontosoria) | 1 / 4 / 11 |
| 258 | **Tapeinidium** (C. Presl) C. Chr. | 7 / 12 / 17 |
| | Wibelia = Tapeinidium | |

## DENNSTAEDTIACEAE (Monachosoroideae)     1(1) / 1(5)

| | | |
|---|---|---|
| 251 | **Monachosorum** Hance | 0 / 1 / 5 |

## Dicksoniaceae see Cyatheaceae

## DIPTERIDACEAE     1(5) / 1(8)

| | | |
|---|---|---|
| 152 | **Dipteris** Reinw. | 1 / 5 / 8 |

Dryopteridaceae see Aspleniaceae (Dryopteridoideae)

Drynariaceae see Polypodiaceae (Drynarioideae)

Elaphoglossaceae see Aspleniaceae (Elaphoglossoideae)

## EQUISETACEAE                                          1(3) / 1(29)

| 011 | Equisetum L. | 1 / 3 / 29 |

## GLEICHENIACEAE                                        4(38) / 4(180)

| 147 | Dicranopteris Bernh. | 1 / 1 / 10 |
| 145 | Diplopterygium (Diels) Nakai | 0 / 1 / 20 |
| 144 | Gleichenia Sm. | 11 / 31 / 50 |
| | Hicriopteris see Dicranopteris, Diplopterygium | |
| 146 | Sticherus C. Presl | 0 / 5 / 100 |

## GRAMMITIDACEAE                                        9(121) / 11(500)

| 214 | Acrosorus Copel. | 0 / 1 / 5 |
| 224 | Anarthropteris Copel. | 1 / 1 / 1 |
| 213 | Calymmodon C. Presl. | 1 / 6 / 25 |
| 212 | Ctenopteris Blume ex Kunze | 6 / 18 / 200 |
| 210 | Grammitis SW. | 4 / 76 / 150 |
| 223 | Loxogramme (Blume) C. Presl | 3 / 6 / 35 |
| 219 | Nematopteris Alderw. | no record |
| | Oreogrammitis = Grammitis | |
| 216 | Prosaptia C. Presl | 1 / 4 / 20 |
| 220 | Scleroglossum Alderw. (Grammitis) | 1 / 3 / 6 |
| 211 | Xiphopteris Kaulf. | 0 / 6 / 50 |

Hemionitidaceae see Adiantaceae (Adiantoideae)

## HYMENOPHYLLACEAE                                      18(100) / 33(465)

| 122 | Abrodictyum C. Presl | 0 / 1 / 1 |
| 115 | Amphipterum (Copel.) Copel. | 0 / 4 / 4 |
| 128 | Callistopteris Copel. (Trichomanes) | 0 / 2 / 5 |
| 127 | Cephalomanes C. Presl. | 2 / 5 / 10 |
| 119 | Crepidomanes (C. Presl) C. Presl (Trichomanes) | 0 / 5 / 20 |
| | Crepidopteris see Crepidomanes | |
| 124 | Gonocormus Bosch (Trichomanes) | 0 / 3 / 6 |
| 106 | Hymenophyllum Sm. | 12 / 14 / 25 |
| | Macroglena = Selenodesmium | |
| 115 | Mecodium C. Presl ex Copel. (Hymenophyllum) | 0 / 5 / 100 |
| 108 | Meringium C. Presl (Hymenophyllum) | 0 / 16 / 60 |
| 131 | Microgonium C. Presl | 0 / 10 / 12 |
| 106 | Microtrichomanes (Mett.) Copel. | 3 / 3 / 14 |
| 110 | Myriodon (Copel.) Copel. | 0 / 2 / 2    endemic in New Guinea |
| 129 | Nesopteris Copel. (Trichomanes) | 2 / 2 / 4 |

| 123 | **Pleuromanes** C. Presl (Trichomanes) | 2 / 2 / 3 |
| 125 | **Selenodesmium** (Prantl) Copel. (Trichomanes) | 2 / 2 / 10 |
| 111 | **Sphaerocionium** C. Presl (Hymenophyllum) | 0 / 1 / 70 |
| 117 | **Trichomanes** L. | 12 / 12 / 25 |
| 117 | **Vandenboschia** Copel. (Trichomanes) | 9 / 9 / 25 |

## ISOETACEAE
1(2) / 2(77)

| 009 | **Isoetes** L. | 0 / 2 / 75 |

Lindsaeaceae see Dennstaedtiaceae (Lindsaeoideae)

Loxogrammaceae see Grammitidaceae

## LYCOPODIACEAE
1(41) / 5(450)

| 003 | **Lycopodium** L. | 20 / 41 / 450 |
| | Urostachys = Lycopodium | |

## MARATTIACEAE
3(30) / 7(100)

| 019 | **Angiopteris** Hoffm. | 2 / 10 / 10 |
| 025 | **Christensenia** Maxon | 1 / 1 / 1 |
| 023 | Macroglossum Copel. | no record |
| 022 | **Marattia** Sw. | 1 / 19 / 60 |

## MARSILEACEAE
1(1) / 3(72)

| 410 | **Marsilea** L. | 0 / 1 / 65 |

## MATONIACEAE
1(1) / 2(4)

| 149 | Matonia R. Br. | no record |
| 150 | **Phanerosorus** Copel. | 0 / 1 / 2 |

Monachosoraceae see Dennstaedticaea (Monachosoroideae)

Oleandraceae see Davalliaceae (Oleandrioideae)

## OPHIOGLOSSACEAE
3(11) / 4(65)

| 012 | **Botrychium** Sw. | 0 / 1 / 23 |
| 015 | **Helminthostachys** Kaulf. | 1 / 1 / 1 |
| | Ophioderma = Ophioglossum | |
| 016 | **Ophioglossum** L. | 2 / 9 / 40 |

## OSMUNDACEAE
2(4) / 3(18)

| 030 | **Leptopteris** C. Presl. | 3 / 3 / 7 |
| | Osmunda see Helminthostachys (Ophioglossaceae) | |
| 029 | **Todea** Willd. ex Bernh. | 0 / 1 / 1 |

## PARKERIACEAE
1(1) / 1(4)

| 037 | **Ceratopteris** Brongn. | 1 / 1 / 4 |

## PLAGIOGYRIACEAE
1(4) / 1(37)

| 031 | **Plagiogyria** (Kunze) Mett. | 1 / 4 / 37 |

## PLATYZOMATACEAE                          0 / 1(1)
038   Platyzoma R. Br.                       no record

## POLYPODIACEAE (Drynarioideae)           6(12) / 9(40)
156   **Aglaomorpha** Schott                 2 / 1 / 10
153   **Drynaria** (Bory) J. Sm.             3 / 7 / 20
159   **Drynariopsis** (Copel.) Ching        1 / 1 / 1
        Dryostachyum = Aglaomorpha
158   **Holostachyum** (Copel.) Ching (Aglaomorpha)   0 / 1 / 1      endemic in New Guinea
155   **Merinthosorus** Copel.               1 / 2 / 2
        Photinopteris see Merinthosorus
161   **Thayeria** Copel.                    0 / 1 / 1

## POLYPODIACEAE (Microsorioideae)         11(60) / 19(180)
183   Arthromeris (T. Moore) J. Sm.          no record
176   Christopteris Copel. (Christiopteris)  no record
170   **Colysis** C. Presl.                  1 / 1 / 30
181   **Crypsinus** C. Presl.                1 / 16 / 40
167   **Dendroconche** Copel.                0 / 2 / 2      endemic in New Guinea
        Dendroglossa = Colysis
172   Diblemma J. Sm.                        no record
178   **Grammatopteridium** Alderw.          0 / 1 / 2
180   **Holcosorus** T. Moore                0 / 1 / 3
169   **Lecanopteris** Reinw.                1 / 3 / 3
        Lepisorus see Crypsinus, Microsorum
173   **Leptochilus** Kaulf.                 0 / 1 / 1
166   **Microsorum** Link (Microsorium, Polypodium)   12 / 29 / 60
        Myrmecophila = Lecanopteris
179   **Oleandropsis** Copel.                0 / 1 / 1      endemic in New Guinea
        Phymatodes = Microsorum
        Pleopeltis see Crypsinus, Microsorum
171   **Podosorus** Holttum                  0 / 1 / 1
184   Polypodiopteris C. Reed                no record
177   Pycnoloma C. Chr.                      no record
182   **Selliguea** Bory                     0 / 4 / 5

## POLYPODIACEAE (Platycerioideae)         2(18) / 4(120)
        Cyclophorus = Pyrrosia
        Drymoglossum = Pyrrosia
        Niphobolus = Pyrrosia
162   **Platycerium** Desv.                  0 / 2 / 17
        Pteropsis = Pyrrosia
163   **Pyrrosia** Mirbel                    7 / 16 / 100

## POLYPODIACEAE (Pleopeltoideae)          3(9) / 14(70)
196   **Belvisia** Mirbel                    2 / 7 / 15

|     | Hymenolepis = Belvisia | |
| --- | --- | --- |
| 193 | **Lemmaphyllum** C. Presl. | 1 / 1 / 5 |
| 198 | Neocheiropteris Christ | no record |
| 195 | **Paragramma** (Blume) T. Moore | 0 / 1 / 2 |
|     | Weatherbya = Lemmaphyllum | |

## POLYPODIACEAE (Polypodioideae)   3(78) / 13(140)

|     |     |     |
| --- | --- | --- |
| 203 | Dictymia J. Sm. | no record |
| 207 | **Goniophlebium** C. Presl. | 2 / 7 / 20 |
| 206 | **Polypodium** L. | 12 / 70 / 75 |
| 208 | **Thylacopteris** Kunze ex J. Sm. | 0 / 1 / 2 |

## PSILOTACEAE   2(7) / 2(12)

|     |     |     |
| --- | --- | --- |
| 001 | **Psilotum** Sw. | 2 / 4 / 4 |
| 002 | **Tmesipteris** Bernh. | 3 / 3 / 6 |

Pteridaceae see Adiantaceae (Pteridoideae)

## SALVINIACEAE   1(1) / 1(12)

|     |     |     |
| --- | --- | --- |
| 413 | **Salvinia** Séguier | 0 / 1 / 12 |

## SCHIZAEACEAE   2(18) / 4(150)

|     |     |     |
| --- | --- | --- |
|     | Lophidium = Schizaea | |
| 034 | **Lygodium** Sw. | 6 / 10 / 35 |
| 032 | **Schizaea** Sm. | 2 / 8 / 30 |

## SELAGINELLACEAE   1(54) / 1(700)

|     |     |     |
| --- | --- | --- |
| 008 | **Selaginella** Pal. | 12 / 54 / 700 |

Taenitidaceae see Adiantaceae (Adiantoideae)

Tectariaceae see Aspleniaceae (Tectarioideae)

## THELYPTERIDACEAE   18(261) / 30(900)

|     |     |     |
| --- | --- | --- |
| 261 | **Amauropelta** Kunze | 0 / 1 / 200 |
| 282 | **Ampelopteris** Kunze | 0 / 1 / 1 |
| 287 | **Amphineuron** Holttum | 1 / 6 / 12 |
| 276 | **Chingia** Holttum | 0 / 4 / 17 |
| 286 | **Christella** A. Léveillé | 2 / 11 / 51 |
| 266 | **Coryphopteris** Holttum | 3 / 25 / 48 |
| 270 | **Cyclosorus** Link | 13 / 54 / 80 |
|     | Dictyocline see Stenogramme | |
|     | Goniopteris see Cyclosorus, Dryopteris | |
|     | Lastrea = Thelypteris | |
|     | Leptogramma = Thelypteris | |
| 267 | **Macrothelypteris** (H. Itô) Ching | 0 / 4 / 10 |
|     | Mesochlaena = Sphaerostephanos | |
| 273 | **Mesophlebion** Holttum | 0 / 5 / 17 |

| | | |
|---|---|---|
| 265 | **Parathelypteris** (H. Itô) Ching (Thelypteris) | 0 / 1 / 15 |
| 262 | **Phegopteris** Fée | 0 / 2 / 3 |
| 274 | **Plesioneuron** (Holttum) Holttum (Thelypteris) | 2 / 34 / 50 |
| 285 | **Pneumatopteris** Nakai | 0 / 30 / 80 |
| 272 | **Pronephrium** C. Presl | 0 / 14 / 68 |
| 285 | **Pseudocyclosorus** Ching (Pneumatopteris) | 0 / 1 / 12 |
| 263 | **Pseudophegopteris** Ching | 0 / 3 / 20 |
| 281 | **Sphaerostephanos** J. Sm. | 4 / 58 / 180 |
| 260 | **Thelypteris** Schmidel | 7 / 7 / 8 |

## THYRSOPTERIDACEAE

2(4) / 3(20)

| | | |
|---|---|---|
| 237 | **Cibotium** Kaulf. | 0 / 1 / 12 |
| 236 | **Culcita** C. Presl. | 1 / 3 / 7 |

Tmesipteridaceae see Psilotaceae

Vittariaceae see Adiantaceae (Vittarioideae)

## ACANTHACEAE                           32(129) / 357(4350)

| | | |
|---|---|---|
| Acanthus L. | 2 / 4 / 30 | |
| Ancylacanthus = Ptyssiglottis | | |
| Aphelandra R. Br. | 0 / 1 / 170 | intro. cult. orn. |
| Aporuellia see Ruellia | | |
| Asystasia Blume | 1 / 5 / 70 | intro. cult. orn. |
| Barleria L. | 1 / 1 / 250 | intro.? cult. orn. |
| Beloperone = Justicia | | |
| Calophanoides = Justicia | | |
| Calycacanthus Schumann | 1 / 1 / 1 | endemic in New Guinea |
| Crossandra Salisb. | 1 / 1 / 50 | intro. cult. orn. |
| Dicliptera Juss. | 0 / 5 / 150 | |
| Dipteracanthus = Ruellia | | |
| Eranthemum L. | 4 / 7 / 30 | intro. |
| Gendarussa Nees (Justicia) | 0 / 1 / 2 | |
| Graptophyllum Nees | 1 / 5 / 10 | |
| Gymnophragma Lindau | 0 / 1 / 1 | endemic in New Guinea |
| Hemigraphis Nees (Strobilanthes) | 3 / 20 / 90 | |
| Hulemacanthus S. Moore | 0 / 3 / 3 | endemic in New Guinea |
| Hygrophila R. Br. | 0 / 3 / 100 | |
| Hypoestes Sol. ex R. Br. | 0 / 2 / 40 | |
| Jadunia Lindau | 0 / 2 / 2 | endemic in New Guinea |
| Justicia L. | 4 / 9 / 420 | some intro. orn. |
| Lepidagathis Willd. | 1 / 5 / 100 | |
| Leptosiphonium = Ruellia | | |
| Nelsonia R. Br. | no record | |
| Nothoruellia = Ruellia | | |
| Oreothyrsus Lindau (Ptyssiglottis) | 0 / 2 / 3 | |
| Peristrophe Nees | 0 / 6 / 15 | |
| Phlogacanthus Nees | 0 / 1 / 15 | |
| Pigafetta Adans. = Erianthemum | | |
| Polytrema C.B. Clarke | no record | |
| Psacadocalymma = Stethoma | | |
| Pseuderanthemum Radlk. | 7 / 11 / 60 | |
| Ptyssiglottis T. Anderson | 0 / 1 / 30 | |
| Rhaphidospora Nees (Justicia) | 0 / 2 / 12 | |
| Ruellia L. | 4 / 13 / 150 | |
| Rungia Nees | 1 / 6 / 50 | |
| Sanchezia Ruíz & Pavón | 1 / 1 / 20 | intro. cult. orn. |
| Schaueria see Hyptis (Lamiaceae) | | |
| Staurogyne Wallich | 1 / 3 / 80 | |
| Stethoma Raf. | no record | |
| Strobilanthes Blume | 1 / 4 / 250 | intro. cult. orn. |

**Thunbergia** Retz.     3 / 3 / 100     intro. cult. orn.

## ACTINIDIACEAE     1(89) / 3(355)

**Saurauia** Willd.     8 / 89 / 300

Tremanthera = Saurauia

Aegialitidaceae see Plumbaginaceae

## AGAVACEAE     4(15) / 16(528)

(see also Dracaenaceae, Sansevieraceae)

**Agave** L.     1 / 2 / 300     intro. cult. orn.

Calodracon = Cordyline

**Cordyline** Comm. ex R. Br.     2 / 11 / 15     orn.

Fourcroya = Furcraea

**Furcraea** Vent.     0 / 1 / 20

**Yucca** L.     1 / 1 / 40     intro. cult. orn.

## AIZOACEAE     3(5) / 115(2410)

(see also Molluginaceae)

**Sesuvium** L.     1 / 1 / 7

**Tetragonia** L.     0 / 1 / 60     intro.

**Trianthema** L.     0 / 3 / 9

## ALANGIACEAE     1(9) / 1(17)
(see also Nyssaceae)

**Alangium** Lam.     2 / 9 / 17

## ALISMATACEAE     3(6) / 11(96)

**Caldesia** Parl. (Alisma)     0 / 3 / 4

**Limnophyton** Miq.     0 / 1 / 1

**Sagittaria** L.     0 / 2 / 20

Alliaceae see Liliaceae

## ALSEUOSMIACEAE     0 / 3(8)

Wittsteinia F. Muell.     no record

## AMARANTHACEAE     12(26) / 72(800)

**Achyranthes** L.     2 / 2 / 6     cult. ed.

**Aerva** Forssk.     0 / 1 / 10

**Alternathera** Forssk.     2 / 3 / 80

**Amaranthus** L.     5 / 9 / 60     1 intro. cult. ed.

**Celosia** L.     1 / 1 / 50     intro. cult. orn.

**Cyathula** Blume     1 / 1 / 20

**Deeringia** R. Br.     0 / 3 / 7

Euxolus = Amaranthus

**Gomphrena** L.     1 / 2 / 100     orn.

**Iresine** P. Browne     1 / 1 / 80     intro. cult. orn.

Leiospermum = Psilotrichum

| | | |
|---|---|---|
| **Psilotrichum** Blume | 1 / 1 / 15 | intro. |
| **Ptilotus** R. Br. | 0 / 1 / 100 | |
| **Pupalia** Juss. | 0 / 1 / 4 | ident.? |

Teleianthera = Alternanthera

## AMARYLLIDACEAE

9(19) / 75(1100)

(see also Agavaceae, Hypoxidaceae)

| | | |
|---|---|---|
| **Crinum** L. | 1 / 8 / 130 | |
| **Eucharis** Planchon & Linden | 1 / 1 / 20 | intro. cult. orn. |
| Eurycles = Proiphys | | |
| **Habranthus** Herbert | 0 / 1 / 10 | intro. cult. orn. |
| **Haemanthus** L. | 1 / 1 / 21 | intro. cult. orn. |
| **Hippeastrum** Herbert | 1 / 1 / 76 | intro. cult. orn. |
| **Hymenocallis** Salisb. | 0 / 2 / 40 | intro. cult. orn. |
| **Pancratium** L. | 0 / 1 / 20 | intro. cult. orn. |
| **Proiphys** Herbert | 1 / 2 / 3 | |
| **Zephyranthes** Herbert | 2 / 2 / 71 | intro. cult. orn. |

## ANACARDIACEAE

15(75) / 73(850)

| | | |
|---|---|---|
| **Anacardium** L. | 1 / 1 / 8 | intro. cult. econ. |
| **Buchanania** Sprengel | 4 / 18 / 25 | |
| **Campnosperma** Thwaites | 2 / 6 / 10 | |
| **Dracontomelon** Blume | 2 / 4 / 8 | |
| Duckera = Rhus | | |
| **Euroschinus** Hook. f. | 0 / 3 / 6 | |
| Evia = Spondias | | |
| **Gluta** L. | 0 / 2 / 30 | |
| **Koordersiodendron** Engl. | 0 / 1 / 1 | |
| Lannea A. Rich. | no record | |
| **Mangifera** L. | 4 / 10 / 35 | |
| Nothopegiopsis = Semecarpus | | |
| Odina = Lannea | | |
| Oncocarpus = Semecarpus | | |
| Parishia Hook. f. | no record | |
| **Pentaspadon** Hook. f. | 2 / 2 / 6 | |
| **Pleiogynium** Engl. | 2 / 2 / 3 | |
| **Rhus** L. | 1 / 6 / 200 | |
| **Semecarpus** L. f. | 5 / 18 / 60 | |
| Skoliostigma = Spondias | | |
| **Spondias** L. | 2 / 2 / 10 | |

## ANCISTROCLADACEAE

0 / 1(12)

Ancistrocladus see Durandea (Linaceae)

| ANNONACEAE | 32(156 / 128(2050) | |
|---|---|---|
| (see also Eupomatiateae) | | |
| **Alphonsea** Hook. f. & Thomson | 0 / 3 / 30 | |
| **Annona** L. | 3 / 4 / 100 | intro. cult. ed. |
| **Artabotrys** R. Br. | 0 / 4 / 100 | |
| Beccariodendron see Goniothalamus | | |
| **Cananga** (DC.) Hook. f. & Thomson | 1 / 1 / 2 | cult. |
| **Cleistopholis** Pierre ex Engl. | 1 / 1 / 4 | intro. |
| **Cyathocalyx** Champ. ex Hook f. & Thomson | 2 / 9 / 38 | |
| **Cyathostemma** Griff. | 0 / 3 / 8 | |
| **Desmos** Lour. | 0 / 25 / 30 | |
| Drepananthus = Cyathocalyx | | |
| **Fissistigma** Griffith | 0 / 3 / 60 | |
| **Friesodielsia** Steenis | 1 / 1 / 55 | |
| **Goniothalamus** (Blume) Hook. f. & Thomson | 2 / 14 / 115 | |
| **Meiogyne** Miq. | 0 / 1 / 10 | |
| **Melodorum** Lour. | 0 / 2 / 4 | |
| **Miliusa** Leschen ex A. DC. | 0 / 1 / 40 | |
| Mitrella = Fissistigma | | |
| **Mitrephora** (Blume) Hook. f. & Thomson | 0 / 3 / 25 | |
| **Oncodostigma** Diels | 0 / 1 / 3 | |
| **Oreomitra** Diels | 0 / 1 / 1 | endemic in New Guinea |
| **Orophea** Blume | 0 / 5 / 60 | |
| Oxymitra = Friesodielsia | | |
| **Papualthia** Diels | 1 / 8 / 20 | |
| **Petalolophus** Schumann | 0 / 1 / 1 | endemic in New Guinea |
| **Phaeanthus** Hook. f. & Thomson | 0 / 3 / 20 | |
| **Platymitra** Boerl. | 0 / 1 / 2 | ident.? |
| **Polyalthia** Blume | 2 / 16 / 120 | |
| **Polyaulax** Backer | 0 / 1 / 1 | |
| **Popowia** Endl. | 1 / 13 / 50 | |
| **Pseuduvaria** Miq. | 0 / 13 / 18 | |
| **Rauwenhoffia** R. Scheffer | 0 / 4 / 5 | |
| **Rollinia** A. St-Hil. | 1 / 1 / 65 | intro. cult. |
| Saccopetalum = Miliusa | | |
| **Schefferomitra** Diels | 0 / 1 / 1 | endemic in New Guinea |
| **Stelechocarpus** (Blume) Hook. f. & Thomson | 0 / 1 / 5 | |
| Unona = Xylopia | | |
| **Uvaria** L. | 2 / 8 / 100 | |
| **Xylopia** L. | 2 / 3 / 100 | |
| **APIACEAE** | 11(21) / 420(3100) | |
| **Apium** L. | 0 / 4 / 20 | cult. |
| **Centella** L. | 1 / 1 / 20 | |

| | | |
|---|---|---|
| **Coriandrum** L. | 1 / 1 / 2 | intro. cult. econ. |
| **Daucus** L. | 1 / 1 / 22 | intro. cult. ed. |
| Didiscus = Trachymene | | |
| **Falcaria** Bernh. (Oenanthe) | 0 / 1 / 5 | |
| **Hydrocotyle** L. | 1 / 3 / 75 | |
| **Oenanthe** L. | 0 / 1 / 30 | |
| **Oreomyrrhis** Endl. | 0 / 6 / 25 | |
| **Petroselinum** Hill | 1 / 1 / 3 | intro. cult. ed. |
| **Sanicula** L. | 0 / 1 / 37 | |
| **Trachymene** Rudge | 0 / 1 / 12 | |

| | | |
|---|---|---|
| # APOCYNACEAE | 34(167) / 215(2100) | |
| **Adenium** Roemer & Schultes | 1 / 1 / 5 | intro. cult. orn. |
| **Allamanda** L. | 4 / 4 / 12 | intro. cult. orn. |
| **Allowoodsonia** Markgraf | 1 / 1 / 1 | endemic in Papuasia |
| **Alstonia** R. Br. | 7 / 16 / 43 | |
| **Alyxia** R. Br. | 4 / 36 / 120 | |
| **Anodendron** A. DC. | 2 / 2 / 20 | |
| Bleekeria = Ochrosia | | |
| **Carissa** L. | 0 / 1 / 37 | |
| **Carruthersia** Seemann | 4 / 4 / 6 | |
| **Catharanthus** G. Don f. | 2 / 2 / 8 | intro. cult. orn. |
| **Cerbera** L. | 2 / 3 / 6 | |
| Chaetosus = Parsonsia | | |
| **Chilocarpus** Blume | 0 / 1 / 15 | |
| Clitandropsis = Melodinus | | |
| **Delphyodon** K. Schum. | 0 / 1 / 1 | endemic in New Guinea |
| Discalyxia = Alyxia | | |
| Ecdysanthera Hook. & Arn. | no record | |
| Ervatamia = Tabernaemontana | | |
| Excavatia = Ochrosia | | |
| Gynopogon = Alyxia | | |
| **Ichnocarpus** R. Br. | 1 / 6 / 18 | |
| **Kentrochrosia** Lauterb. & K. Schum. | 2 / 2 / 3 | |
| Kopsia see Kentrochrosia | | |
| Lactaria = Ochrosia | | |
| Lamechites = Micrechtites | | |
| **Lepinia** Decne. | 1 / 2 / 3 | |
| **Lepiniopsis** Valeton | 0 / 1 / 2 | |
| **Leuconotis** Jack | 0 / 1 / 10 | |
| Lochnera = Catharanthus | | |
| **Lyonsia** R. Br. (Parsonsia) | 0 / 7 / 24 | |
| **Melodinus** Forster & Forster f. | 1 / 9 / 50 | |
| **Micrechites** Miq. | 2 / 3 / 20 | |

| | | |
|---|---|---|
| **Neisosperma** Raf. (Ochrosia) | 2 / 5 / 20 | |
| Neowollastonia = Melodinus | | |
| **Nerium** L. | 1 / 1 / 2 | intro. cult. orn. |
| **Ochrosia** Juss. | 7 / 13 / 23 | |
| Orchipeda = Voacanga | | |
| Pagiantha = Tabernaemontana | | |
| **Papuechites** Markgraf | 0 / 1 / 3 | endemic in New Guinea |
| Paralstonia = Alyxia | | |
| **Parsonsia** R. Br. | 3 / 20 / 80 | |
| **Plumeria** L. | 3 / 3 / 7 | intro. cult. orn. |
| Pseudochrosia = Ochrosia | | |
| Pseudowillughbeia = Melodinus | | |
| **Rauvolfia** L. | 0 / 2 / 110 | |
| Rejoua = Tabernaemontana | | |
| Rhynchodia Benth. | no record | |
| Strophanthus see Papuechites | | |
| **Tabernaemontana** L. | 4 / 10 / 100 | |
| **Thevetia** L. | 1 / 1 / 8 | intro. cult. orn. |
| **Urceola** Roxb. | 0 / 1 / 15 | |
| Vinca see Catharanthus | | |
| **Voacanga** Thouars | 0 / 3 / 12 | |
| **Willughbeia** Roxb. | 1 / 1 / 15 | intro. |
| **Wrigthia** R. Br. | 0 / 3 / 24 | |

## APONOGETONACEAE
1(2) / 1(44)

| | | |
|---|---|---|
| **Aponogeton** L. f. | 0 / 2 / 44 | |

## Apostasiaceae see Orchidaceae

## AQUIFOLIACEAE
2(16) / 4(420)

| | | |
|---|---|---|
| Idenburgia = Sphenostemon | | |
| **Ilex** L. | 2 / 13 / 400 | |
| Nouhuysia = Sphenostemon | | |
| Phelline Labill. | no record | |
| **Sphenostemon** Baillon | 0 / 3 / 7 | |

## ARACEAE
27(155) / 106(2950)

| | | |
|---|---|---|
| **Acorus** L. (considered as seperate family) | 0 / 1 / 2 | |
| **Aglaonema** Schott | 4 / 4 / 21 | intro. cult. orn. |
| **Alocasia** (Schott) G. Don. f. | 6 / 12 / 70 | |
| **Amorphophallus** Blume ex Decne. | 1 / 2 / 90 | |
| **Amydrium** Schott | 0 / 1 / 4 | |
| Anthurium see Pothos | | |
| Arisacontis see Cyrtosperma | | |
| **Caladium** Vent. | 1 / 1 / 7 | intro. cult. orn. |
| **Colocasia** Schott | 2 / 2 / 6 | intro.? cult. ed. |

| | | |
|---|---|---|
| **Cryptocoryne** Fischer ex Wydler | 0 / 3 / 50 | |
| **Cyrtosperma** Griffith | 2 / 11 / 11 | |
| Diandriella = Homalomena | | |
| **Dieffenbachia** Schott | 2 / 2 / 25 | intro. cult. orn. |
| Epipremnopsis = Amydrium | | |
| **Epipremnum** Schott | 3 / 7 / 15 | |
| **Holochlamys** Engl. | 0 / 1 / 2 | endemic in New Guinea |
| **Homalomena** Schott | 4 / 25 / 140 | |
| **Lasia** Lour. | 0 / 1 / 3 | |
| Monstera see Rhaphidophora | | |
| **Philodendron** Schott | 7 / 7 / 500 | intro. cult. orn. |
| Piptospatha N.E. Br. | no record | |
| **Pistia** L. | 1 / 1 / 1 | |
| **Pothos** L. | 6 / 15 / 50 | |
| Raphidophora = Rhaphidophora | | |
| **Rhaphidophora** Hassk. | 6 / 30 / 140 | |
| **Schismatoglottis** Zoll. & Moritzi | 2 / 11 / 100 | |
| Schizocasia see Alocasia | | |
| **Scindapsus** Schott | 3 / 5 / 40 | |
| **Spathiphyllum** Schott | 2 / 3 / 36 | |
| **Syngonium** Schott | 1 / 3 / 33 | intro. cult. orn. |
| **Typhonium** Schott | 0 / 3 / 30 | |
| **Xanthosoma** Schott | 2 / 2 / 45 | intro. cult. ed. |
| Xenophya see Alocasia | | |
| **Zamioculcas** Schott | 1 / 1 / 1 | intro. cult. orn. |
| **Zantedeschia** Sprengel (Richardia) | 1 / 1 / 6 | intro. cult. orn. |

## ARALIACEAE

| | | |
|---|---|---|
| | 15(225) / 58(800) | |
| **Anakasia** Philipson | 0 / 1 / 1 | endemic in New Guinea |
| Anomopanax = Mackinlaya | | |
| **Aralia** L. | 0 / 3 / 36 | |
| **Arthrophyllum** Blume | 0 / 3 / 31 | |
| Boerlagiodendron = Osmoxylon | | |
| Brassaia = Schefflera | | |
| **Cheirodendron** Nutt. ex Seemann | 1 / 1 / 8 | intro.? |
| Cissodendron see Schefflera | | |
| **Delarbrea** Vieill. | 1 / 1 / 4 | |
| **Gastonia** Comm. ex Lam. | 6 / 6 / 10 | |
| Gelibia = Polyscias | | |
| **Harmsiopanax** Warb. | 0 / 2 / 3 | |
| Hedera see Schefflera | | |
| Heptapleurum = Schefflera | | |
| Kissodendron = Polyscias | | |
| **Mackinlaya** F. Muell. | 1 / 3 / 5 | |

| | | |
|---|---|---|
| **Meryta** Forster & Forster f. | 2 / 2 / 16 | |
| Nothopanax = Polyscias | | |
| **Osmoxylon** Miq. | 8 / 10 / 50 | |
| Palmervandenbroekia = Polyscias | | |
| Panax see Polyscias | | |
| Paratropia = Schefflera | | |
| Peekeliopanax = Gastonia | | |
| Pentapanax Seemann | no record | |
| Plerandra = Schefflera | | |
| **Polyscias** Forster & Forster f. | 10 / 23 / 100 | |
| **Schefflera** Forster & Forster f. | 12 / 170 / 800 | |
| Scheffleropsis = Schefflera | | |
| Sciadophyllum = Schefflera | | |
| **Strobilopanax** R. Viguier | no record | |
| Tetraplasandra = Gastonia | | |
| Tieghemopanax = Polyscias | | |
| **Trevesia** Vis. | no record | |

## ARAUCARIACEAE
2(5) / 2(31)

| | |
|---|---|
| **Agathis** Salisb. | 2 / 2 / 13 |
| **Araucaria** Juss. | 3 / 3 / 18 |
| Dammara = Agathis | |

## ARECACEAE
40(314) / 207(2675)

| | | |
|---|---|---|
| Actinophloeus = Ptychosperma | | |
| **Actinorhytis** H. Wendl. & Drude | 2 / 2 / 2 | |
| Adelonenga = Hydriastele | | |
| Archontophoenix H.A. Wendl. & Drude | no record | |
| **Areca** L. | 7 / 14 / 50 | |
| **Arenga** Lab. | 0 / 3 / 17 | |
| Bacularia = Linospadix | | |
| Barkerwebbia = Heterospathe | | |
| **Borassus** L. | 0 / 1 / 7 | |
| **Brassiophoenix** Burret | 0 / 2 / 2 | endemic in New Guinea |
| **Calamus** L. | 3 / 50 / 370 | |
| **Calyptrocalyx** Blume | 0 / 39 / 40 | |
| Carpentaria Becc. | no record | |
| Carpoxylon H. Wendl. & Drude | no record | |
| **Caryota** L. | 1 / 2 / 12 | |
| Chambeyronia Vieill. | no record | |
| **Clinostigma** H. Wendl. | 1 / 2 / 10 | |
| **Cocos** L. | 1 / 1 / 1 | cult. econ. |
| Coleospadix = Drymophloeus | | |
| **Corypha** L. | 0 / 1 / 8 | |
| Cyphophoenix H. Wendl. ex Hook. f. | no record | |

| | |
|---|---|
| **Cyrtostachys** Blume | 1 / 11 / 12 |
| **Daemonorops** Blume | 0 / 1 / 114 |
| Dammera = Licuala | |
| Didymosperma = Arenga | |
| **Drymophloeus** Zipp. | 6 / 6 / 15 |
| **Elaeis** Jacq. | 1 / 1 / 2 |
| Grisebachia see Laccospadix | |
| **Gronophyllum** R. Scheffer | 0 / 7 / 14 |
| **Gulubia** Becc. | 4 / 6 / 10 |
| Gulubiopsis = Gulubia | |
| **Heterospathe** R. Scheffer | 6 / 16 / 20 |
| **Howeia** Becc. | 1 / 1 / 2    cult. orn. |
| **Hydriastele** H. Wendl. & Drude | 0 / 7 / 9 |
| Kajewskia see Veitchia | |
| Kentia = Gronophyllum | |
| **Korthalsia** Blume | 0 / 2 / 25 |
| **Laccospadix** Drude & H. Wendl. | |
| (Calyptrocalyx) | 0 / 1 / 1 |
| Leptophoenix = Nengella | |
| **Licuala** Thunb. | 3 / 36 / 108 |
| **Linospadix** H. Wenl. | 0 / 5 / 11 |
| **Livistona** R. Br. | 1 / 6 / 29 |
| **Metroxylon** Rottb. | 3 / 3 / 8 |
| **Nenga** H. Wendl. & Drude | 0 / 2 / 2 |
| **Nengella** Becc. (Gronophyllum) | 0 / 19 / 19 |
| Nipa = Nypa | |
| **Normanbya** F. Muell. ex Becc. (Ptychosperma) | 0 / 1 / 1 |
| **Nypa** Steck (Nipa) | 1 / 1 / 1 |
| Oncosperma Blume | no record |
| **Orania** Zipp. | 0 / 13 / 17 |
| Paragulubia = Gulubia | |
| Paralinospadix = Calyptrocalyx | |
| **Physokentia** Becc. | 3 / 3 / 6 |
| **Pigafetta** (Blume) Becc. | 0 / 2 / 3 |
| **Pinanga** Blume | 0 / 1 / 120 |
| **Pritchardia** Seemann & H. Wendl. | 1 / 1 / 36 |
| Pseudopinanga = Pinanga | |
| Ptychandra = Heterospathe | |
| **Ptychococcus** Becc. | 0 / 7 / 7    endemic in Papuasia |
| Ptychoraphis see Rhopaloblaste | |
| **Ptychosperma** Labill. | 8 / 27 / 28 |
| Rehderophoenix = Drymophloeus | |
| **Rhopaloblaste** R. Scheffer | 1 / 4 / 7 |
| Schizospatha = Calamus | |

| | | |
|---|---|---|
| Siphokentia Burret | no record | |
| **Sommieria** Becc. | 0 / 3 / 3 | endemic in New Guinea |
| Strongylocaryum = Ptychosperma | | |
| **Veitchia** H. Wendl. | 1 / 1 / 18 | intro. cult. orn. |

## ARISTOLOCHIACEAE

| | |
|---|---|
| **ARISTOLOCHIACEAE** | 1(11) / 7(410) |
| **Aristolochia** L. | 3 / 11 / 300 |

## ASCLEPIADACEAE

| | | |
|---|---|---|
| **ASCLEPIADACEAE** | 28(182) / 348(2900) | |
| **Asclepias** L. | 1 / 1 / 120 | |
| Astelma Schl. = Papuastelma | | |
| **Brachystelma** R. Br. (Microstemma) | 0 / 1 / 60 | |
| **Calotropis** R. Br. | 1 / 1 / 2 | intro. cult. orn. |
| **Ceropegia** L. | 1 / 3 / 160 | intro. cult. orn. |
| Collyris = Dischidia | | |
| **Conchophyllum** Blume | 0 / 1 / 10 | |
| **Cryptostegia** R. Br. | 1 / 1 / 2 | intro. cult. orn. |
| **Cynanchum** L. | 0 / 2 / 100 | |
| **Dischidia** R. Br. | 3 / 30 / 80 | |
| **Finlaysonia** Wallich | 0 / 2 / 2 | |
| Gomphocarpus = Asclepias | | |
| **Gongronema** (Endl.) Decne. | 0 / 2 / 15 | |
| **Gymnanthera** R. Br. | 0 / 1 / 4 | |
| **Gymnema** R. Br. | 0 / 5 / 25 | |
| **Heterostemma** Wight & Arn. | 0 / 5 / 30 | |
| **Hoya** R. Br. | 8 / 79 / 90 | |
| **Ischnostemma** King & Gamble | 0 / 1 / 1 | |
| **Marsdenia** R. Br. | 1 / 18 / 100 | |
| **Papuastelma** Bullock | 0 / 1 / 1 | endemic in New Guinea |
| **Pentatropis** Wight & Arn. | 0 / 1 / 6 | ident.? |
| **Phyllanthera** Blume | 0 / 2 / 2 | |
| **Physostelma** Wight | 0 / 1 / 6 | |
| Pterostelma = Hoya | | |
| **Sarcolobus** R. Br. | 1 / 3 / 4 | |
| **Secamone** R. Br. | 0 / 1 / 100 | |
| **Spathidolepis** Schltr. | 0 / 1 / 1 | endemic in New Guinea |
| Sperlingia = Hoya | | |
| **Stephanotis** Thouars | 0 / 1 / 15 | ident.? |
| **Streptomanes** Schumann ex Schltr. | 0 / 1 / 1 | endemic in New Guinea |
| **Toxocarpus** Wight & Arn. | 0 / 6 / 40 | |
| **Tylophora** R. Br. (Riedelia) | 2 / 10 / 50 | |
| **Vincetoxicum** Wolf | 0 / 1 / 15 | |

Asparagaceae see Liliaceae

## ASTERACEAE                      69(217) / 1317(21000)

| | | |
|---|---|---|
| **Abrotanella** Cass. | 0 / 1 / 16 | |
| **Achillea** L. | 0 / 1 / 85 | |
| **Adenostemma** Forster & G. Forster | 1 / 5 / 20 | intro. |
| **Ageratum** L. | 1 / 2 / 43 | intro. |
| **Anaphalis** DC. | 1 / 5 / 100 | |
| **Arrhenechthites** Mattf. | 0 / 5 / 7 | |
| **Artemisia** L. | 0 / 1 / 300 | |
| Astelma R. Br. see Helichrysum | | |
| **Aster** L. | 0 / 1 / 250 | |
| Baccharis see Pluchea | | |
| **Bidens** L. | 2 / 3 / 233 | |
| **Blumea** DC. | 10 / 18 / 75 | |
| **Brachionostylum** Mattf. | 0 / 1 / 1 | endemic in New Guinea |
| **Brachycome** Cass. | 0 / 2 / 66 | |
| **Centipeda** Lour. | 0 / 2 / 4 | |
| **Chrysanthemum** L. | 0 / 1 / 2 | |
| **Conyza** Less. | 0 / 2 / 50 | intro.? |
| Coreopsis L. | no record | |
| **Cosmos** Cav. | 0 / 1 / 26 | |
| **Cotula** L. | 0 / 4 / 80 | |
| **Crassocephalum** Moench | 1 / 1 / 30 | |
| Crepis L. | no record | |
| **Dahlia** Cav. | 1 / 1 / 28 | intro. cult. orn. |
| **Dichrocephala** L'Hérit. ex DC. | 0 / 2 / 4 | |
| Distreptus = Elephantopus | | |
| **Eclipta** L. | 1 / 2 / 4 | |
| **Elephantopus** L. | 1 / 2 / 32 | intro. |
| **Eleutheranthera** Poit. ex Bosc. | 1 / 1 / 1 | |
| **Emilia** Cass. | 1 / 2 / 24 | |
| **Epaltes** Cass. | 0 / 1 / 15 | |
| **Erechtites** Raf. | 2 / 5 / 5 | nat. |
| **Erigeron** L. | 1 / 5 / 200 | intro. |
| **Ethulia** L. f. | 0 / 1 / 1 | |
| Felicia Cass. | no record | |
| **Galinsoga** Ruíz & Pavón | 0 / 1 / 14 | |
| Glossogyne see Bidens | | |
| **Gnaphalium** L. | 0 / 5 / 150 | |
| **Gynura** Cass. | 0 / 2 / 50 | |
| Harrisonia Necker = Xeranthemum | | |
| Hecatactis = Keysseria | | |
| **Helianthus** L. | 1 / 1 / 67 | intro. cult. orn. ed |
| **Helichrysum** Miller | 0 / 1 / 500 | |
| **Ischnea** F. Muell. | 0 / 5 / 5 | endemic in New Guinea |

| | | |
|---|---|---|
| **Ixeris** (Cass.) Cass. (Lactuca) | 0 / 2 / 50 | |
| **Jungia** L. f. (Trinacte) | no record | |
| **Keysseria** Lauterb. | 1 / 9 / 15 | |
| **Lactuca** L. | 1 / 7 / 100 | intro. cult. orn. |
| Lagenifera = Lagenophora | | |
| **Lagenophora** Cass. | 0 / 3 / 9 | |
| **Melanthera** J.P. Rohr | 1 / 1 / 20 | |
| **Microglossa** DC. | 1 / 1 / 60 | |
| **Mikania** Willd. | 2 / 2 / 300 | |
| **Myriactis** Less. | 0 / 3 / 12 | |
| **Olearia** Moench | 0 / 19 / 130 | |
| **Papuacalia** Veldk. | 0 / 10 / 10 | endemic in New Guinea |
| **Phacellothrix** F. Muell. | 0 / 1 / 1 | |
| **Piora** J. Koster | 0 / 1 / 1 | endemic in New Guinea |
| **Pluchea** Cass. | 0 / 1 / 40 | |
| **Pseudelephantopus** Rohr (Elephantopus) | 0 / 1 / 2 | |
| **Pterocaulon** Elliott | 0 / 2 / 18 | |
| **Raoulia** Hook. f. ex Raoul (Gnaphalium) | 0 / 1 / 20 | |
| **Rhamphogyne** S. Moore | 0 / 1 / 2 | |
| **Senecio** L. | 1 / 6 / 1500 | |
| **Sigesbeckia** L. | 0 / 1 / 9 | |
| **Sonchus** L. | 0 / 2 / 62 | |
| **Sphaeranthus** L. | 0 / 2 / 38 | |
| **Spilanthes** Jacq. | 1 / 1 / 6 | |
| **Struchium** P. Browne | 0 / 1 / 1 | |
| **Synedrella** Gaertner | 1 / 1 / 2 | |
| **Tagetes** L. | 1 / 1 / 50 | intro. cult. orn. |
| **Taraxacum** Wigg. | 0 / 1 / 60 | |
| **Tetramolopium** Nees | 0 / 18 / 32 | |
| **Tithonia** Desf. ex Juss. | 0 / 1 / 10 | |
| **Tridax** L. | 1 / 1 / 26 | nat. |
| **Verbesina** L. | 0 / 2 / 150 | |
| **Vernonia** Schreber | 2 / 9 / 1000 | |
| **Vittadinia** A. Rich. | 0 / 1 / 15 | |
| **Wedelia** Jacq. (Wollastonia) | 3 / 9 / 70 | |
| Wollastonia DC. ex Decne. (Wedelia) | no record | |
| Xanthium L. | no record | |
| Xeranthemum L. | no record | |
| **Youngia** Cass. | 1 / 1 / 28 | |
| **Zinnia** L. | 1 / 1 / 22 | intro. cult. orn. |

## AUSTROBAILEYACEAE   1(1) / 1(1)

| | |
|---|---|
| **Austrobaileya** C. White | 0 / 1 / 1 |

**AVICENNIACEAE**  1(4) / 1(14)
Avicennia L.  4 / 4 / 14

**BALANOPACEAE**  0 / 1(9)
Balanops Baillon  no record
Trilocularia = Balanops

**BALANOPHORACEAE**  2(6) / 18(44)
**Balanophora** Forster & G. Forster  0 / 5 / 16
**Langsdorffia** C. Martius  0 / 1 / 3

**BALSAMINACEAE**  1(6) / 2(850)
**Impatiens** L.  2 / 6 / 850  nat.

Barclayaceae see Nymphaeaceae

**BARRINGTONIACEAE**  5(34) / 20(280)
**Barringtonia** Forster & G. Forster  13 / 30 / 39
Careya see Planchonia, Barringtonia
**Cariniana** Casar.  1 / 1 / 15  intro. cult.
Chydenanthus Miers  no record
Eschweileria see Osmoxylon (Araliaceae)
**Gustavia** L.  1 / 1 / 4  intro.
**Planchonia** Blume  1 / 2 / 5

**BASELLACEAE**  2(2) / 4(15)
**Andredera** Juss.  0 / 1 / 2  intro. orn.
**Basella** L.  0 / 1 / 5  cult. ed.

**BATACEAE**  1(1) / 1(2)
**Batis** P. Browne  0 / 1 / 2

Batidaceae see Bataceae

**BEGONIACEAE**  1(87) / 2(900)
**Begonia** L.  4 / 87 / 900
Symbegonia = Begonia

**BIGNONIACEAE**  12(39) / 112(725)
Bignonia see Dolichandrone
Campsis Lour.  no record
**Crescentia** L.  0 / 1 / 6  cult. orn.
**Deplanchea** Vieill.  0 / 3 / 5
Diplanthera = Deplanchea
**Dolichandrone** (Fenzl) Seemann  1 / 2 / 9
**Jacaranda** Juss.  1 / 1 / 30  intro. cult. orn.
**Lamiodendron** Steenis (Fernandoa)  0 / 1 / 1  endemic in New Guinea
**Neosepicaea** Diels  0 / 4 / 4
Nyctocalos Teijsm. & Binnend.  no record

| | | |
|---|---|---|
| **Pandorea** (Engl.) Spach | 1 / 8 / 8 | |
| **Pyrostegia** C. Presl | 1 / 1 / 4 | intro. cult. orn. |
| Radermachera Zoll. & Moritzi | no record | |
| **Spathodea** Pal. | 1 / 1 / 1 | intro. cult. orn. |
| **Tabebuia** Gomes ex DC. | 1 / 1 / 100 | intro. |
| **Tecoma** Juss. | 1 / 2 / 12 | intro.? cult. orn. |
| **Tecomanthe** Baillon | 0 / 14 / 5 | |

Bischofiaceae see Euphorbiaceae

## BIXACEAE

| | | |
|---|---|---|
| **BIXACEAE** | 2(2) / 3(16) | |
| **Bixa** L. | 1 / 1 / 1 | nat. cult. econ. |
| **Cochlospermum** Kunth | 0 / 1 / 12 | |

## BOMBACACEAE

| | | |
|---|---|---|
| **BOMBACACEAE** | 6(7) / 30(250) | |
| Bombacopsis = Pachira | | |
| **Bombax** L. | 2 / 2 / 8 | |
| **Camptostemon** Masters | 0 / 1 / 2 | |
| **Ceiba** Miller | 1 / 1 / 4 | intro. cult. |
| Cumingia = Camptostemon | | |
| **Durio** Adans. | 1 / 1 / 27 | intro. cult. ed. |
| Eriodendron = Ceiba | | |
| Gossampinus see Bombax | | |
| **Ochroma** SW. | 1 / 1 / 1 | intro. cult. econ. |
| **Pachira** Aublet | 1 / 1 / 24 | intro. cult. |
| Papuodendron see Hibiscus (Malvaceae) | | |
| Salmalia = Bombax | | |

Bonnetiaceae see Theaceae

## BORAGINACEAE

| | | |
|---|---|---|
| **BORAGINACEAE** | 12(46) / 156(2500) | |
| **Argusia** Boehmer (Tournefortia) | 1 / 1 / 3 | |
| Carmona = Ehretia | | |
| **Coldenia** L. | 0 / 1 / 1 | |
| **Cordia** L. | 3 / 5 / 250 | some cult. |
| **Crucicaryum** Brand | 0 / 1 / 1 | endemic in New Guinea |
| **Cynoglossum** L. | 0 / 2 / 55 | |
| Echinospermum = Lappula | | |
| **Ehretia** P. Browne | 1 / 3 / 50 | |
| Havilandia = Trigonotis | | |
| **Heliotropium** L. | 2 / 5 / 250 | |
| Lappula Gilib. | no record | |
| **Lithospermum** L. (Trigonotis) | 0 / 1 / 59 | |
| Messerschmidia = Argusia | | |
| **Myosotis** L. | 0 / 2 / 50 | |
| Plagiobothrys see Trigonotis | | |

Rhabdia = Rotula
Rotula Lour.                              no record
Tiaridium = Heliotropium
**Tournefortia** L.                       1 / 14 / 150
**Trichodesma** R. Br.                    0 / 1 / 35
**Trigonotis** Steven                     0 / 10 / 50
Zoelleria = Trigonotis

## BRASSICACEAE                          7(21) / 390(3000)
**Brassica** L.                           2 / 2 / 30            intro. cult. ed.
**Capsella** Medikus                      0 / 1 / 5
**Cardamine** L.                          1 / 4 / 130
**Lepidium** L.                           0 / 3 / 150
**Nasturtium** R. Br. (Rorippa)           1 / 5 / 6
Papuzilla = Lepidium
**Raphanus** L.                           1 / 1 / 8            intro. cult. ed.
**Rorippa** Scop.                         0 / 3 / 70

## BROMELIACEAE                         1(1) / 46(2110)
**Ananas** Miller                         1 / 1 / 8            nat. cult. ed.

## BURMANNIACEAE                        4(14) / 21(165)
**Burmannia** L.                          1 / 6 / 57
**Gymnosiphon** Blume                     0 / 6 / 29
**Scaphiophora** Schltr.                  0 / 1 / 2
**Thismia** Griffith                      0 / 1 / 24

## BURSERACEAE                          8(55) / 18(540)
**Aucoumea** Pierre                       1 / 1 / 1            intro. cult.
Bursera see Protium
**Canarium** L.                           12 / 28 / 75         some cult. ed.
**Dacryodes** Vahl                        0 / 1 / 40
**Garuga** Roxb.                          1 / 1 / 4
**Haplolobus** H.J. Lam.                  3 / 19 / 22
**Protium** Burm. f.                      0 / 2 / 85
**Santiria** Blume                        0 / 2 / 24
**Scutinanthe** Thwaites                  0 / 1 / 2

Butomaceae see Limnocharitaceae

## BYBLIDACEAE                          1(2) / 2(4)
**Byblis** Salisb.                        0 / 2 / 2

## CABOMBACEAE                          1(1) / 2(8)
**Cabomba** Aublet                        0 / 1 / 7

## CAESALPINIACEAE                      22(107) / 162(2000)
Afzelia see Intsia

| | | |
|---|---|---|
| **Amherstia** Wallich | 0 / 1 / 1 | intro. cult. orn. |
| **Bauhinia** L. | 3 / 7 / 250 | intro. cult. orn. |
| **Brownea** Jacq. | 0 / 1 / 30 | intro. cult. orn. |
| **Caesalpinia** L. | 5 / 8 / 100 | |
| **Cassia** L. | 10 / 25 / 535 | |
| **Crudia** Schreber | 2 / 7 / 55 | |
| **Cynometra** L. | 2 / 12 / 70 | |
| Dansera = Dialium | | |
| **Delonix** Raf. | 1 / 1 / 10 | intro. cult. orn. |
| Dialium L. | no record | |
| **Gigasiphon** Drake (Bauhinia) | 0 / 1 / 5 | |
| Gleditsia L. | no record | |
| Guilandina = Caesalpinia | | |
| Hardwickia Roxb. | no record | |
| **Intsia** Thouars | 1 / 3 / 3 | |
| Kalappia Kosterm. | no record | |
| **Kingiodendron** Harms (Oxystigma) | 3 / 5 / 6 | |
| **Koompassia** Maingay | 0 / 1 / 3 | |
| **Lasiobema** (Korth.) Miq. (Bauhinia) | 0 / 1 / 10 | intro. cult. |
| **Lysiphyllum** (Benth.) De Wit (Bauhinia) | 0 / 1 / 7 | |
| Macrolobium see Intsia | | |
| **Maniltoa** R. Scheffer | 2 / 21 / 25 | |
| **Mezonevron** Desf. (Caesalpinia) | 2 / 3 / 30 | |
| Outea = see Intsia | | |
| Pahudia = Intsia | | |
| **Parkinsonia** L. | 1 / 1 / 19 | intro. cult. orn. |
| **Peltophorum** (Vogel) Benth. | 0 / 1 / 8 | intro. cult. orn. |
| **Phanera** Lour. | 0 / 2 / 60 | |
| Poinciana = Caesalpinia | | |
| Pseudocynometra = Maniltoa | | |
| **Saraca** L. | 0 / 3 / 9 | |
| Schizoscyphus = Maniltoa | | |
| Schizosiphon = Maniltoa | | |
| Senna = Cassia | | |
| **Tamarindus** L. | 1 / 1 / 1 | intro.? cult. |
| Trachylobium Hayne (Hymenaea) | no record | |
| **Tylosema** (Schweinf.) Torre & Hillcoat | 0 / 1 / 4 | intro. cult. |

## CALLITRICHACEAE

1(1) / 1(17)

| | |
|---|---|
| **Callitriche** L. | 0 / 1 / 17 |

## CAMPANULACEAE

6(10) / 80(1000)

(see also Lobeliaceae)

Campanumoea = Codonopsis

| | | |
|---|---|---|
| **Codonopsis** Wallich | 0 / 1 / 30 | |
| Isotoma = Solenopsis | | |
| Laurentia = Solenopsis | | |
| **Peracarpa** Hook. f. & Thomson | 0 / 1 / 1 | |
| Phyllocharis = Ruthiella | | |
| **Ruthiella** Steenis | 0 / 4 / 4 | endemic in New Guinea |
| **Sclerotheca** A. DC. | 1 / 1 / 3 | |
| **Solenopsis** C. Presl | 0 / 1 / 25 | |
| **Wahlenbergia** Schrader ex Roth | 0 / 2 / 150 | |

Candolleaceae see Stylidiaceae

## CANNABIDACEAE

| | | |
|---|---|---|
| | 1(1) / 2(3) | |
| **Cannabis** L. | 0 / 1 / 1 | cult. econ. |

## CANNACEAE

| | | |
|---|---|---|
| | 1(2) / 1(25) | |
| **Canna** L. | 2 / 2 / 25 | intro. cult. orn. |

## CAPPARACEAE

| | | |
|---|---|---|
| | 4(18) / 45(675) | |
| **Cadaba** Forssk. | 0 / 1 / 30 | |
| **Capparis** L. | 2 / 11 / 250 | |
| **Cleome** L. | 1 / 4 / 150 | intro. |
| **Crateva** L. (Crataeva) | 1 / 2 / 6 | |
| Gynandropsis = Cleome | | |
| Polanisia = Cleome | | |

Capparidaceae see Capparaceae

## CAPRIFOLIACEAE

| | | |
|---|---|---|
| | 3(4) / 16(365) | |
| **Lonicera** L. | 1 / 1 / 180 | intro. cult. orn. |
| **Sambucus** L. | 0 / 2 / 20 | |
| **Viburnum** L. | 0 / 1 / 150 | |

## CARDIOPTERIDACEAE

| | | |
|---|---|---|
| | 1(3) / 1(2) | |
| Aspidocarya = Cardiopteris | | |
| **Cardiopteris** Wallich ex Royle | 0 / 2 / 2 | |
| Peripterygim = Cardiopteris | | |

## CARICACEAE

| | | |
|---|---|---|
| | 1(1) / 4(31) | |
| **Carica** L. | 1 / 1 / 22 | cult. ed. |

## CARYOPHYLLACEAE

| | | |
|---|---|---|
| | 7(15) / 89(2070) | |
| **Cerastium** L. | 0 / 2 / 60 | |
| **Drymaria** Willd. ex Schultes | 0 / 1 / 50 | |
| **Polycarpaea** Lam. | 0 / 2 / 50 | |
| **Sagina** L. | 0 / 5 / 25 | |
| **Scleranthus** L. | 0 / 1 / 10 | |
| **Spergula** L. | 0 / 1 / 5 | intro. |

Spergularia (Pers.) Pers. & C. Presl — no record
Stellaria L. — 1 / 3 / 120

## CASUARINACEAE — 3(14) / 4(100)
Casuarina L. — 2 / 8 / 70
Ceuthostoma L. Johnson — 0 / 1 / 2
Gymnostoma L. Johnson — 0 / 5 / 20

## CECROPIACEAE — 1(4) / 6(200)
Poikilospermum Zipp. ex Miq. — 0 / 4 / 20

## CELASTRACEAE — 17(37) / 94(1300)
Bhesa Buch.-Ham. ex Arn. — 0 / 1 / 5
Brassiantha A.C. Sm. — 0 / 1 / 1 — endemic in New Guinea
Cassine L. — 0 / 1 / 80
Celastrus L. — 0 / 3 / 31
Combretopsis = Lophopyxis
Elaeodendron = Cassine
Euonymus L. — 0 / 2 / 177
Gymnosporia = Maytenus
Hippocratea L. — 0 / 4 / 120
Kurrimia = Bhesa
Loesneriella A.C. Sm. — 2 / 3 / 26
Lophopetalum Wight ex Arn. — 0 / 4 / 18
Lophopyxis Hook f. — 1 / 1 / 1
Maytenus Molina — 0 / 3 / 225
Microtropis Wallich ex Meissner — no record
Perrottetia Kunth — 1 / 1 / 15
Pleurostylia Wight & Arn. — 0 / 1 / 6
Reissantia Hallé — 0 / 1 / 7
Salacia L. — 6 / 6 / 150
Salacicratea Loes. (Salacia) — 0 / 1 / 1
Siphonodon Griffith — 0 / 3 / 7
Solenospermum = Lophopetalum
Xylonymus Kalkman ex Ding Hou — 0 / 1 / 1 — endemic in New Guinea

## CENTROLEPIDACEAE — 2(4) / 3(28)
Centrolepis Labill. — 0 / 3 / 20
Gaimardia Gaudich. — 0 / 1 / 2

## CERATOPHYLLACEAE — 1(2) / 1(2)
Ceratophyllum L. — 1 / 2 / 2

CHailletiaceae see Dichapetalaceae

## CHENOPODIACEAE — 5(8) / 120(1300)
Arthrocnemon Moq. (Arthrocnemum) see Halosarcia

| | | |
|---|---|---|
| Bassia All. | no record | |
| **Chenopodium** L. | 0 / 4 / 150 | intro. |
| Halocnemium = Tecticornia | | |
| **Halosarcia** P.G. Wilson | 0 / 1 / 23 | |
| **Salicornia** L. | 0 / 1 / 13 | ident.? |
| **Salsola** L. | 0 / 1 / 150 | |
| Suaeda Forssk. ex Scop. | no record | |
| **Tecticornia** Hook. f. | 0 / 1 / 3 | |

## CHLORANTHACEAE — 3(6) / 4(58)

| | |
|---|---|
| **Ascarina** Foster & Foster f. | 2 / 4 / 8 |
| **Chloranthus** Sw. | 0 / 1 / 15 |
| **Sarcandra** Gardner | 0 / 1 / 3 |

## CHRYSOBALANACEAE — 5(13) / 17(460)

| | | |
|---|---|---|
| Angelesia = Licania | | |
| **Cyclandrophora** Hassk. (Atuna) | 0 / 1 / 7 | |
| **Hunga** Pancher ex Prance | 0 / 4 / 11 | |
| **Licania** Aublet | 0 / 1 / 171 | |
| Maranthes see Parinari | | |
| **Parastemon** A. DC. | 0 / 1 / 2 | ident.? |
| **Parinari** Aublet | 4 / 6 / 44 | |

## CLETHRACEAE — 1(6) / 1(64)

| | |
|---|---|
| **Clethra** L. | 0 / 6 / 64 |

## CLUSIACEAE — 7(121) / 47(1350)

| | | |
|---|---|---|
| **Calophyllum** L. | 11 / 34 / 187 | |
| Calysaccion see Mammea | | |
| Cratoxylum Blume | no record | |
| **Garcinia** L. | 12 / 68 / 200 | 1 intro. cult. |
| **Hypericum** L. | 0 / 8 / 370 | |
| **Kayea** Wall (Mesua) | 0 / 1 / 35 | |
| Lolanara = Mammea | | |
| **Mammea** L. | 4 / 3 / 50 | |
| **Mesua** L. | 0 / 1 / 40 | |
| Ochrocarpus = Mammea | | |
| **Pentaphalangium** Warb. (Garcinia) | 2 / 5 / 7 | |
| Tripetalum = Garcinia | | |
| Xanthochymus = Garcinia | | |

Cochlospermaceae see Bixaxeae

## COMBRETACEAE — 4(39) / 20(500)

| | |
|---|---|
| **Combretum** Loefl. | 0 / 4 / 250 |
| Laguncularia see Lumnitzera | |
| **Lumnitzera** Willd. | 1 / 2 / 2 |

Pyrrhanthus = Lumnitzera
Quisqualis L. ............................ 0 / 1 / 16
Terminalia L. ........................... 17 / 32 / 150

## COMMELINACEAE ............. 8(36) / 42(620)
Aclisia = Pollia
Amischotolype Hassk. .............. 0 / 2 / 20
Aneilema R. Br. ...................... 1 / 12 / 60
Commelina L. .......................... 4 / 6 / 150
Cyanotis D. Don ...................... 1 / 3 / 30
Floscopa Lour. ........................ 0 / 1 / 20
Forrestia = Amischotolype
Murdannia Royle ..................... 1 / 3 / 50
Pollia Thunb. .......................... 3 / 8 / 17
Rhoeo = Tradescantia
Tradescantia L. ....................... 0 / 1 / 65          intro. cult. orn.
Zebrina = Tradescantia

## Compositae see Asteraceae

## CONNARACEAE ................. 2(15) / 20(380)
Connarus L. ............................ 3 / 12 / 100
Rourea Aublet .......................... 2 / 3 / 85
Santaloides = Rourea

## CONVOLVULACEAE ........... 15(92) / 58(1650)
Aniseia Choisy ........................ 1 / 1 / 5
Bonamia Thouars ..................... 0 / 1 / 45
Calonyction = Ipomoea
Calystegia R. Br. ..................... no record
Convolvulus Juss. .................... no record
Cuscuta L. .............................. 0 / 2 / 145
Dichondra Forster & Forster f. .. 0 / 1 / 9
Erycibe Roxb. ......................... 1 / 15 / 67
Evolvulus L. ........................... 0 / 2 / 98
Hewittia Wight & Arn. ............. 0 / 1 / 1
Ipomoea L. ............................. 14 / 37 / 500     some nat. cult. ed.
Jacquemontia Choisy ............... 0 / 5 / 120
Lepistemon Blume .................... 1 / 3 / 10
Merremia Dennst. ex Endl. ....... 5 / 14 / 70
Mina Llave & Lex. ................... 0 / 1 / 1           cult. orn.
Operculina A. Silva Manso ....... 1 / 5 / 15
Pharbitis = Ipomoea
Porana Burm. f. ....................... 0 / 2 / 20
Quamoclit = Ipomoea
Stictocardia Hallier f. .............. 0 / 2 / 12

## CORIARIACEAE                          1(1) / 1(5)
**Coriaria** L.                          1 / 1 / 5

## CORNACEAE                             1(3) / 12(90)
**Mastixia** Blume                       1 / 3 / 13

## CORSIACEAE                            1(23) / 2(26)
**Corsia** Becc.                         2 / 23 / 25

## CORYNOCARPACEAE                       1(1) / 1(4)
**Corynocarpus** Forster & Forster f.    1 / 1 / 4

Costaceae see Zingiberaceae

## CRASSULACEAE                          1(1) / 33(1280)
**Bryophyllum** = Kalanchoe
**Kalanchoe** Adans.                     1 / 1 / 125                    nat.

Cruciferae see Brassicaceae

## CRYPTERONIACEAE                       2(2) / 5(11)
**Crypteronia** Blume                    0 / 1 / 4
**Dactylocladus** Oliver                 0 / 1 / 1

Ctenolophonaceae see Linaceae

## CUCURBITACEAE                         20(56) / 121(735)
**Alsomitra** (Blume) M. Roemer          0 / 5 / 2
**Benincasa** Savi                       1 / 1 / 1              intro. cult. ed.
Bryonia see Melothria
Bryonopsis = Kedrostis
Cerasiocarpum = Kedrostis
**Citrullus** Schrader                   1 / 1 / 3             intro. cult. ed.
**Coccinia** Wight & Arn.                1 / 1 / 30                    nat.
**Cucumis** L.                           2 / 2 / 30            intro. cult. ed.
**Cucurbita** L.                         3 / 3 / 27            intro. cult. ed.
**Diplocyclos** (Endl.) Post & Kuntze    1 / 1 / 5                cult. orn.
**Gymnopetalum** Arn.                    0 / 1 / 3
**Gynostemma** Blume                     0 / 2 / 2
**Kedrostis** Medikus                    1 / 1 / 23
**Lagenaria** Ser.                       2 / 2 / 6             intro. cult. ed.
**Luffa** Miller                         3 / 3 / 6         2 intro. cult. econ.
Macrozanonia = Alsomitra
**Melothria** L.                         2 / 14 / 10
**Momordica** L.                         1 / 4 / 45                    nat.
**Mukia** Arn.                           0 / 2 / 4
**Neoalsomitra** Hutch.                  0 / 2 / 12
Rhynchocarpa = Kedrostis

| | | |
|---|---|---|
| **Sechium** P. Browne | 1 / 1 / 6 | intro. cult. |
| **Trichosanthes** L. | 1 / 6 / 15 | nat. cult. ed. |
| **Zanonia** L. | 0 / 1 / 1 | |
| **Zehneria** Endl. | 1 / 3 / 30 | |

## CUNONIACEAE    10(48) / 24(340)

| | | |
|---|---|---|
| Ackama = Caldcluvia | | |
| **Acsmithia** Hoogl. | 0 / 3 / 14 | |
| **Aistopetalum** Schltr. | 0 / 2 / 2 | endemic in New Guinea |
| Betchea = Caldcluvia | | |
| **Caldcluvia** D. Don | 2 / 5 / 11 | |
| **Ceratopetalum** Sm. | 0 / 2 / 5 | |
| Cremnobates = Schizomeria | | |
| Cunonia see Caldcluvia | | |
| **Geissois** Labill. | 1 / 2 / 18 | |
| **Gillbeea** F. Muell. | 0 / 1 / 2 | |
| Opocunonia = Caldcluvia | | |
| **Pullea** Schltr. | 0 / 3 / 3 | |
| **Schizomeria** D. Don | 3 / 12 / 15 | |
| **Spiraeanthemum** A. Gray | 2 / 5 / 6 | |
| Spiraeopsis = Caldcluvia | | |
| Stollaea = Caldcluvia | | |
| **Weinmannia** L. | 4 / 13 / 190 | |

## CUPRESSACEAE    2(2) / 17(113)

| | | |
|---|---|---|
| **Callitris** Vent. | 0 / 1 / 14 | intro. cult. orn. |
| **Libocedrus** Endl. | 0 / 1 / 8 | |
| Papuacedrus = Libocedrus | | |

Cuscutaceae see Convolvulaceae

## CYCADACEAE    1(2) / 1(20)

| | |
|---|---|
| **Cycas** L. | 1 / 2 / 20 |

## CYMODOCEACEAE    2(3) / 5(16)

| | |
|---|---|
| **Cymodocea** König | 2 / 2 / 4 |
| **Halodule** Endl. | 0 / 1 / 6 |
| Syringodium Kütz. | no record |

## CYPERACEAE    29(243) / 115(3600)

| | |
|---|---|
| Baumea see Machaerina | |
| **Bulbostylis** Kunth (Fimbristylis) | 0 / 2 / 100 |
| Capitularia = Chorizandra | |
| Capitularina = Chorizandra | |
| **Carex** L. | 2 / 36 / 1000 |
| **Carpha** Banks & Sol. ex R. Br. | 0 / 1 / 11 |
| Cephaloscirpus = Mapania | |

Chaetospora see Cladium, Rhynchospora, Schoenus
Chlorocyperus = Cyperus
**Chorizandra** R. Br.                           2 / 1 / 5
**Cladium** P. Browne                            1 / 1 / 2
**Costularia** C.B. Clarke ex Dyer               0 / 1 / 20
Cyclocampe see Lophoschoenus, Schoenus
**Cyperus** L.                                   19 / 53 / 600
Diplacrum = Scleria
Duval-Jouvea = Cyperus
**Eleocharis** R. Br.                            4 / 13 / 150
**Exocarya** Benth.                              0 / 1 / 1
**Fimbristylis** Vahl                            7 / 36 / 150
**Fuirena** Rottb.                               1 / 1 / 30
**Gahnia** Forster & Forster f.                  0 / 3 / 30
Heleocharis = Eleocharis
Helothrix = Schoenus
**Hypolytrum** Rich.                             1 / 2 / 80
Isolepis = Scirpus
Kyllinga = Cyperus
Lampocarya = Gahnia
**Lepidosperma** Labill.                         0 / 1 / 40
**Lepironia** Pers.                              0 / 1 / 1
**Lipocarpha** R. Br.                            0 / 2 / 15
**Lophoschoenus** Stapf (Costularia)             0 / 1 / 11
**Machaerina** Vahl                              1 / 9 / 45
**Mapania** Aublet                               4 / 18 / 50
Mariscus = Cyperus
**Oreobolus** R. Br.                             0 / 2 / 8
Paramapania = Mapania
**Pycreus** Pal. (Cyperus)                       1 / 1 / 70
**Remirea** Aublet (Cyperus)                     0 / 1 / 1
**Rhynchospora** Vahl                            1 / 10 / 200
**Schoenus** L.                                  1 / 9 / 80
**Scirpodendron** Zipp. ex Kurz                  1 / 1 / 1
**Scirpus** L.                                   0 / 13 / 200
**Scleria** P. Bergius                           7 / 20 / 200
Thoracostachyum = Mapania
Torulinium = Cyperus
**Tricostularia** Nees ex Lehm.                  0 / 1 / 5
**Uncinia** Pers.                                0 / 2 / 35

# DAPHNIPHYLLACEAE                                1(19 / 1(10)
**Daphniphyllum** Blume                          1 / 1 / 10

**DATISCACEAE**  2(2) / 3(4)
Octomeles Miq.  1 / 1 / 1
Tetrameles R. Br.  0 / 1 / 1

**DICHAPETALACEAE**  1(5) / 3(180)
Chailletia = Dichapetalum
Dichapetalum Thouars  3 / 5 / 150
Pentastira = Dichapetalum

**DILLENIACEAE**  3(21) / 12(300)
Dillenia L.  6 / 16 / 60
Hibbertia Andrews  0 / 2 / 122
Tetracera L.  0 / 3 / 40
Wormia = Dillenia

**DIOSCOREACEAE**  1(9) / 8(630)
Dioscorea L.  8 / 9 / 600  cult. ed.

**DIPTEROCARPACEAE**  3(17) / 16(530)
Anisoptera Korth.  0 / 3 / 11
Hopea Roxb.  0 / 13 / 102
Shorea see Hopea
Vateria L.  no record
Vatica L.  0 / 1 / 65

**DRACAENACEAE**  1(4) / 1(40)
Dracaena Vand. ex L.  3 / 4 / 40  1 intro. cult. orn.
Pleomele = Dracaena

**DROSERACEAE**  1(7) / 4(85)
Aldrovanda L.  no record
Drosera L.  0 / 7 / 80

**EBENACEAE**  1(35) / 2(485)
Cargillia = Diospyros
Diospyros L.  13 / 35 / 475
Maba = Diospyros

Ehretiaceae see Boraginaceae

**ELAEAGNACEAE**  1(1) / 3(45)
Elaeagnus L.  0 / 1 / 40

**ELAEOCARPACEAE**  7(121) / 11(220)
Aceratium DC.  2 / 14 / 20
Anoniodes see Sloanea
Antholoma see Sloanea
Aristotelia L'Hérit. (Sericolea)  0 / 1 / 5
Dubouzetia Pancher ex Brongn. & Gris  0 / 4 / 10

| | | |
|---|---|---|
| **Echinocarpus** Blume (Sloanea) | 0 / 1 / 10 | |
| **Elaeocarpus** L. | 16 / 68 / 100 | |
| Hormopetalum = Sericolea | | |
| Mischopleura see Sericolea | | |
| Phoenicosperma = Sloanea | | |
| Pyrsonota = Sericolea | | |
| **Sericolea** Schltr. | 0 / 15 / 15 | endemic in New Guinea |
| **Sloanea** L. | 2 / 18 / 100 | |

## ELATINACEAE
2(2) / 2(32)

| | | |
|---|---|---|
| **Bergia** L. | 0 / 1 / 20 | |
| **Elatine** L. | 0 / 1 / 12 | |

## EPACRIDACEAE
4(33) / 31(400)

| | | |
|---|---|---|
| **Decatoca** F. Muell. | 0 / 1 / 1 | endemic in New Guinea |
| Epacris Cav. | no record | |
| **Leucopogon** R. Br. (Styphelia) | 0 / 2 / 150 | |
| **Styphelia** Sm. | 0 / 23 / 25 | |
| **Trochocarpa** R. Br. | 0 / 7 / 12 | |

## ERICACEAE
6(436) / 103(3350)

| | | |
|---|---|---|
| **Agapetes** D. Don ex G. Don. f. | 0 / 11 / 95 | |
| **Dimorphanthera** (Drude) F. Muell. ex J.J. Sm. | 0 / 62 / 67 | |
| **Diplycosia** Blume | 0 / 22 / 97 | |
| Disiphon = Vaccinium | | |
| **Gaultheria** L. | 0 / 11 / 150 | |
| Neojunguhnia = Vaccinium | | |
| Paphia = Agapetes | | |
| **Rhododendron** L. | 4 / 155 / 850 | some cult. orn. |
| **Vaccinium** L. | 1 / 175 / 450 | |

## ERIOCAULACEAE
1(22) / 14(1200)

| | |
|---|---|
| **Eriocaulon** L. | 0 / 22 / 400 |

## ERYTHROXYLACEAE
1(3) / 4(260)

| | |
|---|---|
| **Erythroxylum** P. Browne | 2 / 3 / 250 |

## Escalloniaceae see Grossulariaceae

## EUPHORBIACEAE
52(426) / 326(7750)

(see also Pandaceae, Stilaginaceae)

| | |
|---|---|
| **Acalypha** L. | 6 / 18 / 430 |
| **Actephila** Blume | 1 / 4 / 35 |
| Adelia see Mallotus, Spatiostemon | |
| Adisca = Mallotus | |
| **Agrostistachys** Dalz. | 0 / 2 / 9 |
| **Alchornea** Sw. | 0 / 1 / 70 |

| | | |
|---|---|---|
| Alcinaeanthus = Neoscortechina | | |
| **Aleurites** Forster & G. Forster | 1 / 1 / 6 | |
| **Alphandia** Baillon | 0 / 1 / 3 | |
| **Annesijoa** Pax & K. Hoffm. | 0 / 1 / 1 | endemic in New Guinea |
| **Aporusa** Blume (Aporosa) | 2 / 17 / 75 | |
| Austrobuxus see Kairothamnus | | |
| **Baccaurea** Lour. | 3 / 6 / 80 | |
| Baloghia see Fonatainea | | |
| Bischofia Blume | no record | |
| **Blumeodendron** Kurz | 0 / 2 / 6 | |
| **Breynia** Forster & Forster f. | 2 / 9 / 25 | |
| **Bridelia** Willd. | 2 / 6 / 60 | |
| Carumbium see Homalanthus, Excoecaria | | |
| **Choriceras** Baillon (Dissilaria) | 0 / 1 / 1 | |
| **Claoxylon** A. Juss. | 7 / 29 / 80 | |
| Clarorivinia = Ptychopyxis | | |
| **Cleidion** Blume | 3 / 3 / 25 | |
| **Cleistanthus** Hook. f. ex Planchon | 1 / 9 / 130 | |
| Coccoceras = Mallotus | | |
| Coccoglochidion = Glochidion | | |
| **Codiaeum** A. Juss. | 1 / 5 / 6 | orn. |
| Coelodiscus = Mallotus | | |
| **Croton** L. | 4 / 21 / 750 | |
| **Dimorphocalyx** Thwaites | 0 / 1 / 12 | |
| Dissiliaria see Choriceras | | |
| **Drypetes** Vahl | 6 / 9 / 200 | |
| **Endospermum** Benth. | 5 / 5 / 13 | 1 intro. cult. |
| **Euphorbia** L. | 11 / 14 / 1600 | 1 intro. cult. orn. |
| **Excoecaria** L. | 2 / 3 / 40 | |
| **Fahrenheitia** Reichb. f. & Zoll. | | |
| ex Muell. Arg. (Ostodes) | 0 / 1 / 4 | |
| Flueggea see Securinega | | |
| Flueggeopsis = Phyllanthus | | |
| **Fontainea** Heckel | 0 / 1 / 2 | |
| Gelonium = Suregada | | |
| **Glochidion** Forster & Forster f. | 16 / 77 / 300 | |
| Hemicyclia = Drypetes | | |
| Hemiglochidion = Glochidion | | |
| **Hevea** Aublet | 0 / 1 / 9 | intro. cult. econ. |
| **Homalanthus** A. Juss. | 5 / 12 / 35 | |
| **Homonoia** Lour. | 0 / 1 / 2 | |
| Jatropha see Aleurites, Manihot | | |
| **Kairothamnus** Airy Shaw | 1 / 1 / 1 | endemic in New Guinea |
| **Leptopus** Decne. | 0 / 1 / 20 | |

Longetia see Kairothamnus

**Macaranga** Thouars      22 / 75 / 240

**Mallotus** Lour.      7 / 20 / 140

**Manihot** Miller      1 / 2 / 98      nat. cult. ed.

Mappa = Macaranga

**Margaritaria** L. f.      0 / 1 / 14

**Melanolepis** Reichb. f. ex Zoll.      1 / 1 / 2

Neomphalea see Omphalea

**Neoscortechinia** Pax      1 / 2 / 4

**Octospermum** Airy Shaw      0 / 1 / 1      endemic in New Guinea

**Omphalea** L.      1 / 1 / 20

Pedilanthus Necker ex Poit. (Euphorbia)      no record

**Petalostigma** F. Muell.      0 / 1 / 7

**Phyllanthus** L.      12 / 42 / 600

**Pimelodendron** Hassk.      1 / 1 / 8

**Ptychopyxis** Miq.      0 / 2 / 13

Putranjiva = Drypetes

**Ricinus** L.      1 / 1 / 1      nat. cult.

Sapium see Excoecaria

**Sauropus** Blume      1 / 4 / 40      intro. cult.

Schistostigma = Cleistanthus

Scortechinia = Neoscortechinia

**Sebastiania** Sprengel      1 / 1 / 100

**Securinega** Comm. ex Juss.      2 / 2 / 20

**Spathiostemon** Blume      0 / 1 / 3

Stillingia see Excoecaria

**Suregada** Roxb. ex Rottler      0 / 1 / 40

**Syndyophyllum** Schumann & Lauterb.      0 / 1 / 1

Synostemon see Sauropus

Tetraglochidion = Glochidion

**Trigonostemon** Blume      0 / 2 / 45

**Wetria** Baillon      0 / 1 / 1

## EUPOMATIACEAE      1(1) / 1(2)

**Eupomatia** R. Br.      0 / 1 / 2

## FABACEAE      75(280) / 437(11300)

**Abrus** Adans.      0 / 2 / 17

**Aeschynomene** L.      0 / 3 / 150

**Afgekia** Craib      0 / 1 / 3      intro. cult.

**Aganope** Miq.      1 / 1 / 6      intro.

**Alysicarpus** Desv.      1 / 2 / 25      cult.

Amphicarpaea Elliott ex Nutt.      no record

**Arachis** L.      1 / 1 / 22      intro. cult. ed.

**Atylosia** Wight & Arn.      0 / 2 / 35

| | | |
|---|---|---|
| **Butea** Roxb. ex Willd. | 0 / 1 / 4 | intro. cult. orn. |
| **Cajanus** DC. | 1 / 1 / 2 | intro. cult. ed. |
| **Calopogonium** Desv. | 2 / 2 / 8 | some intro. cult. |
| **Canavalia** DC. | 6 / 6 / 51 | |
| **Castanospermum** A. Cunn. ex Hook. | 1 / 1 / 1 | intro.? cult. |
| **Centrosema** (DC.) Benth. | 2 / 2 / 45 | intro. cult. |
| **Christia** Moench | 0 / 1 / 12 | |
| **Clitoria** L. | 1 / 2 / 70 | intro. |
| **Crotalaria** L. | 5 / 25 / 600 | |
| Cullen see Psoralea | | |
| Cytisus see Pongamia | | |
| **Dalbergia** L. f. | 1 / 11 / 100 | |
| Dendrolobium (Wight & Arn.) Benth. see Desmodium | | |
| **Derris** Lour. | 4 / 19 / 40 | |
| **Desmodium** Desv. | 15 / 40 / 300 | some nat. |
| Dicerma see Desmodium | | |
| **Dioclea** Kunth. | 1 / 1 / 30 | |
| Dolichos see Canavalia, Lablab, Vigna | | |
| **Dumasia** DC. | 0 / 1 / 8 | |
| **Dunbaria** Wight & Arn. | 0 / 3 / 15 | |
| **Eriosema** (DC.) G. Don f. | 0 / 1 / 20 | |
| **Erythrina** L. | 4 / 5 / 108 | 1 intro. cult. orn. |
| **Flemingia** Roxb. ex Aiton & Aiton f. | 1 / 5 / 30 | nat. |
| Gajanus = Inocarpus | | |
| **Galactia** P. Browne | 0 / 1 / 50 | intro. |
| **Gliricidia** Kunth | 1 / 1 / 4 | intro. cult. |
| **Glycine** Willd. | 0 / 2 / 9 | intro. cult. ed. |
| **Gompholobium** Sm. | 0 / 1 / 25 | |
| Hanslia = Desmodium | | |
| Hardenbergia see Vandasia | | |
| **Indigofera** L. | 1 / 10 / 700 | nat. |
| **Inocarpus** Forster & Forster f. | 1 / 3 / 3 | |
| Kennedia Vent. | no record | |
| Kennedya (Kennedia) = Vandasina | | |
| Kiesera = Tephrosia | | |
| **Kunstleria** Prain. | 0 / 1 / 8 | |
| **Lablab** Adans. (Dolichos) | 0 / 1 / 1 | |
| **Lathyrus** L. | 0 / 1 / 150 | intro. |
| **Lespedeza** Michaux | 0 / 1 / 40 | |
| **Lonchocarpus** Kunth | 0 / 1 / 150 | inro. cult. |
| **Lotononis** (DC.) Ecklon & Zeyher | 0 / 1 / 100 | intro. |
| **Lotus** L. | 0 / 1 / 100 | ident.? |
| Lourea = Christia | | |
| **Lupinus** L. | 0 / 4 / 200 | intro. cult. |

| | | |
|---|---|---|
| Macropsychanthus Harms ex | | |
|         Schumann & Lauterb. (Dioclea) | 1 / 2 / 4 | |
| Macroptilium (Benth.) Urban | 2 / 2 / 12 | intro. cult. |
| Macrotyloma (Wight & Arn.) Verdc. | 0 / 1 / 24 | intro. |
| Mastersia Benth. | no record | |
| Maughania J. St-Hil. | no record | |
| Medicago L. | 1 / 1 / 56 | intro. cult. |
| Millettia Wight & Arn. | 1 / 2 / 90 | |
| Moghania = Flemingia | | |
| Monarthrocarpus = Desmodium | | |
| Mucuna Adans. | 7 / 25 / 100 | |
| Mundulea (DC.) Benth. | 0 / 1 / 31 | intro. cult. |
| Neonotonia Lackey | 0 / 1 / 2 | |
| Neustanthus = Pueraria | | |
| Ormocarpum Pal. | 1 / 1 / 20 | ed. |
| Ormosia Jackson | 1 / 1 / 100 | |
| Ostyocarpus Hook. f. | no record | |
| Oxyrhynchus Brandegee | 1 / 1 / 4 | |
| Pachyrhizus Rich. ex DC. | 0 / 1 / 6 | intro. cult. ed. |
| Papilionopsis = Desmodium | | |
| Peekelia = Oxyrhynchus | | |
| Pericopsis Thwaites | 1 / 1 / 4 | |
| Phaseolus L. | 5 / 5 / 50 | intro. cult. ed. |
| Phylacium Bennett | 1 / 1 / 2 | |
| Phyllodium = Desmodium | | |
| Pisum L. | 0 / 1 / 5 | intro. cult. ed. |
| Pongamia Vent. | 1 / 2 / 2 | |
| Psophocarpus Necker ex DC. | 1 / 2 / 9 | intro. cult. ed. |
| Psoralea L. | 0 / 1 / 20 | |
| Pterocarpus Jacq. | 1 / 1 / 20 | |
| Pueraria DC. | 3 / 3 / 20 | |
| Pycnospora R. Br. ex Wight & Arn. | 0 / 1 / 1 | |
| Rhynchosia Lour. | 1 / 2 / 200 | |
| Sesbania Scop. | 1 / 4 / 50 | intro. cult. |
| Shuteria Wight & Arn. | 0 / 1 / 5 | ident.? |
| Smithia Aiton | 0 / 2 / 30 | |
| Sophora L. | 1 / 1 / 52 | |
| Soya = Glycine | | |
| Stizolobium = Mucuna | | |
| Strongylodon Vogel | 1 / 4 / 20 | |
| Stylosanthes Sw. | 1 / 2 / 25 | intro. cult. |
| Tephrosia Pers. | 3 / 12 / 400 | |
| Teramnus P. Browne | 0 / 1 / 8 | intro.? |
| Tipuana (Benth.) Benth. | 0 / 1 / 1 | intro. cult. |

| | |
|---|---|
| **Trifolium** L. | 0 / 3 / 238 |
| **Uraria** Desv. | 2 / 2 / 20 |
| **Vandasina** Rauschert (Vandasia) | 1 / 1 / 1 |
| **Vicia** L. | 0 / 2 / 140 |
| **Vigna** Savi | 4 / 11 / 150 |
| **Zornia** J. Gmelin | 0 / 6 / 75 |

Vicia L. — intro.
Vigna Savi — 2 intro. cult. ed.

## FAGACEAE
3(24) / 7(1050)

| | |
|---|---|
| **Castanopsis** (D. Don) Spach | 0 / 1 / 110 |
| Cyclobalanopsis see Lithocarpus | |
| **Lithocarpus** Blume | 1 / 10 / 300 |
| **Nothofagus** Blume | 0 / 13 / 35 |
| Pasania = Lithocarpus | |
| Quercus see Lithocarpus | |

## FLACOURTIACEAE
16(71) / 88(875)

| | |
|---|---|
| **Baileyoxylon** C. White | 0 / 1 / 1 |
| Bennettia = Bennettiodendron | |
| **Bennettiodendron** Merr. | 0 / 1 / 3 |
| **Casearia** Jacq. | 5 / 34 / 180 |
| **Dovyalis** E. Meyer ex Arn. | 0 / 1 / 15 |
| **Erythrospermum** Lam. | 1 / 1 / 4 |
| **Flacourtia** Comm. ex L'Hérit. | 2 / 5 / 15 |
| Gertrudia = Ryparosa | |
| Gestroa = Erythrospermum | |
| **Homalium** Jacq. | 2 / 14 / 180 |
| **Hydnocarpus** Gaertner | 0 / 1 / 40 |
| **Itoa** Hemsley | 0 / 1 / 2 |
| Leucocorema = Trichadenia | |
| **Osmelia** Thwaites | 1 / 1 / 4 |
| **Pangium** Reinw. | 1 / 1 / 1 |
| **Pseudosmelia** Sleumer | 0 / 1 / 1 |
| **Ryparosa** Blume | 0 / 2 / 18 |
| **Scolopia** Schreber | 0 / 2 / 37 |
| **Trichadenia** Thwaites | 1 / 1 / 2 |
| **Xylosma** Forster f. | 1 / 4 / 85 |

Pangium Reinw. — ed.
Pseudosmelia Sleumer — intro.?

## FLAGELLARIACEAE
1(2) / 1(4)

(see also Hanguanaceae, Joinvilleaceae)

| | |
|---|---|
| **Flagellaria** L. | 2 / 2 / 4 |

Flindersiaceae see Rutaceae

## GENTIANACEAE                              4(27) / 74(1200)

(see also Menyanthaceae)

| | |
|---|---|
| **Cotylanthera** Blume | 0 / 2 / 4 |
| **Exacum** L. | 0 / 1 / 25 |
| **Gentiana** L. | 0 / 21 / 300 |
| Limnanthemum = Nymphoides (Menyanthaceae) | |
| **Swertia** L. | 0 / 3 / 50 |

## GERANIACEAE                               1(11) / 14(730)

| | |
|---|---|
| **Geranium** L. | 0 / 11 / 300 |

## GESNERIACEAE                              13(208) / 146(2400)

| | |
|---|---|
| **Aeschynanthus** Jack | 0 / 39 / 100 |
| Agalmyla Blume | no record |
| **Boea** Comm. ex Lam. | 4 / 9 / 17 |
| **Coronanthera** Vieill. ex C.B. Clarke | 1 / 1 / 11 |
| **Cyrtandra** Forster & Forster f. | 10 / 121 / 350 |
| Cyrtandropsis = Tetraphyllum | |
| **Dichrotrichum** Reinw. ex Vriese (Aeschynanthus) | 0 / 22 / 35 |
| **Didissandra** C.B. Clarke | 0 / 1 / 30 |
| Didymocarpus see Boea | |
| **Epithema** Blume | 1 / 2 / 10 |
| Euthamnus see Aeschynanthus | |
| Isanthera = Rhynchotechum | |
| Loxotis = Rhynchoglossum | |
| **Monophyllaea** R. Br. | 0 / 4 / 20 |
| Oxychlamys = Aeschynanthus | |
| **Rhynchoglossum** Blume | 0 / 2 / 13 |
| **Rhynchotoechum** Blume (Rhynchotechum) | 0 / 2 / 12 |
| **Sepikea** Schltr. | 0 / 1 / 1    endemic in New Guinea |
| **Stauranthera** Benth. | 0 / 1 / 10 |
| **Tetraphyllum** Griffith ex C.B. Clarke | 0 / 2 / 2 |
| Trichosporum = Aeschynanthus | |

## GNETACEAE                                 1(1) / 1(28)

| | |
|---|---|
| **Gnetum** L. | 3 / 1 / 28 |
| Thoa = Gnetum | |

Gonystylaceae see Thymelaeaceae

## GOODENIACEAE                              5(14) / 14(430)

| | |
|---|---|
| **Calogyne** R. Br. | 0 / 1 / 9 |
| **Goodenia** Sm. | 0 / 1 / 170 |
| **Lechenaultia** R. Br. | 0 / 1 / 20 |
| Leschenaultia = Lechenaultia | |

| | |
|---|---|
| **Scaevola** L. | 4 / 10 / 130 |
| **Velleia** Sm. | 0 / 1 / 20 |

Gramineae see Poaceae

## GROSSULARIACEAE     2(38) / 22(325)
Argyrocalymma = Carpodetus
| | |
|---|---|
| **Carpodetus** Forster & Forster f. | 1 / 10 / 10 |
| **Polyosma** Blume | 2 / 28 / 60 |

## GUNNERACEAE     1(2) / 1(40)
| | |
|---|---|
| **Gunnera** L. | 1 / 2 / 40 |

Guttiferae see Clusiaceae

Gyrocarpaceae see Hernandiaceae

## HAEMODORACEAE     1(1) / 16(85)
| | |
|---|---|
| **Haemodorum** Sm. | 0 / 1 / 20 |

## HALORAGIDACEAE     2(11) / 9(120)

(see also Gunneraceae)
Goniocarpus = Haloragis
| | |
|---|---|
| **Haloragis** Forster & Forster f. | 1 / 7 / 26 |
| **Myriophyllum** L. | 0 / 4 / 40 |

## HAMAMELIDACEAE     2(2) / 28(90)
Distyliopsis see Sycopsis
| | |
|---|---|
| **Liquidambar** L. (Altingia) | 0 / 1 / 4 |
| **Sycopsis** Oliver (Distyliopsis) | 0 / 1 / 7 |

## HANGUANACEAE     1(1) / 1(2)
| | |
|---|---|
| **Hanguana** Blume | 1 / 1 / 2 |
| Susum = Hanguana | |

## HELICONIACEAE     1(3) / 1(100)
Folium see Heliconia
| | |
|---|---|
| **Heliconia** L. | 3 / 3 / 100 |
| Heliconiopsis = Heliconia | |

## HERNANDIACEAE     3(10) / 4(68)
| | |
|---|---|
| **Gyrocarpus** Jacq. | 1 / 1 / 3 |
| **Hernandia** L. | 7 / 7 / 24 |
| **Illigera** Blume | 0 / 2 / 18 |

## HIMANTANDRACEAE     1(1) / 1(3)
| | |
|---|---|
| **Galbulimima** Bailey | 0 / 1 / 3 |
| Himantandra = Galbulimima | |

## HYDRANGEACEAE
1(5) / 17(170)

Dichroa Lour.
0 / 5 / 13

## HYDROCHARITACEAE
8(14) / 16(100)

Blyxa Noronha ex Thouars
0 / 4 / 12

Enhalus Rich.
1 / 1 / 1

Halophila Thouars
2 / 4 / 9

Hydrilla Rich.
0 / 1 / 1

Hydrocharis L.
0 / 1 / 2

Ottelia Pers.
1 / 1 / 40

Schizotheca = Thalassia

Thalassia Banks & Sol. ex C. Koenig
1 / 1 / 2

Vallisneria L.
0 / 1 / 8

Hydrocotylaceae see Apiaceae

Hypericaceae see Clusiaceae

## HYPOXIDACEAE
2(6) / 7(120)

Curculigo Gaertner
3 / 7 / 15

Hypoxis L.
0 / 2 / 80

## ICACINACEAE
15(49) / 60(320)

Chariessa = Citronella

Citronella D. Don
1 / 2 / 21

Gastrolepis Tieghem
no record

Gomphandra Wallich ex Lindley
1 / 8 / 33

Gonocaryum Miq.
0 / 2 / 10

Hartleya Sleumer
0 / 1 / 1        endemic in New Guinea

Ioedes Blume
0 / 5 / 28

Lasianthera see Gonocaryum, Medusanthera

Medusanthera Seemann
3 / 3 / 5

Merrilliodendron Kaneh.
1 / 1 / 1

Peekeliodendron = Merrilliodendron

Phytocrene Wallich
0 / 3 / 11

Platea Blume
0 / 2 / 5

Pocillaria = Rhyticaryum

Polyporandra Becc.
1 / 1 / 1

Pseudobotrys Moser
0 / 2 / 2        endemic in New Guinea

Rhyticaryum Becc.
0 / 12 / 12

Stemonurus Blume
4 / 4 / 12

Tylecarpus = Medusanthera

Urandra Thwaites (Stemonurus)
0 / 2 / 17

Villaresia = Citronella

Whitmorea Sleumer
1 / 1 / 1        endemic in Papuasia

## IRIDACEAE 5(7) / 92(1850)

| | | |
|---|---|---|
| Gladiolus L. | 0 / 1 / 180 | intro. cult. orn. |
| Libertia Sprengel (Sisyrinchum) | 0 / 1 / 11 | |
| Patersonia R. Br. | 0 / 1 / 13 | |
| Sisyrinchium L. | 0 / 2 / 100 | intro. cult. |
| Tritonia Ker-Gawler | 0 / 2 / 28 | intro. cult. |

## IXONANTHACEAE 1(2) / 1(2)

| | |
|---|---|
| Ixonanthes Exell & Mendonça | 0 / 2 / 3 |

## JOINVILLEACEAE 1(3) / 1(3)

| | |
|---|---|
| Joinvillea Gaudich. ex Brongn. & Gris | 3 / 3 / 3 |

## JUGLANDACEAE 1(2) / 7(59)

| | |
|---|---|
| Engelhardia Leschen. ex Blume (Engelhardtia) | 0 / 2 / 5 |

## JUNCACEAE 2(6) / 10(325)

| | |
|---|---|
| Juncus L. | 0 / 4 / 225 |
| Luzula DC. | 0 / 2 / 80 |

## JUNCAGINACEAE 1(1) / 4(18)

| | |
|---|---|
| Triglochin L. | 0 / 1 / 14 |

Labiatae see Lamiaceae

## LAMIACEAE 23(59) / 224(5600)

| | | |
|---|---|---|
| Achyrospermum Blume | no record | |
| Acrocephalus Benth. | 0 / 1 / 130 | |
| Ajuga L. | 0 / 1 / 50 | |
| Anisomeles R. Br. | 0 / 2 / 6 | |
| Ballota see Hyptis | | |
| Basilicum Moench | 0 / 1 / 7 | |
| Bysteropogon see Hyptis | | |
| Calamintha see Satureja | | |
| Ceratanthus F. Muell. ex G. Taylor | 0 / 1 / 10 | |
| Coleus = Solenostemon | | |
| Cymaria Benth. | 0 / 2 / 2 | |
| Dysophylla Blume (Pogostemon) | 0 / 2 / 25 | |
| Epimeredi Adans. | no record | |
| Hyptis Jacq. | 1 / 5 / 400 | |
| Leucas R. Br. | 2 / 7 / 150 | |
| Mentha L. | 1 / 1 / 25 | intro. cult. |
| Mesona Blume | 0 / 1 / 3 | |
| Microtoena Prain | 0 / 1 / 25 | |
| Moschosma Reichb. (Basilicum) | 2 / 2 / 7 | |
| Ocimum L. | 3 / 6 / 150 | 2 intro. cult. |
| Orthosiphon Benth. | 1 / 1 / 40 | intro. cult. orn. |

| | | |
|---|---|---|
| Plectranthus L'Hérit. | 1 / 5 / 300 | |
| Pogostemon Desf. | 1 / 7 / 71 | |
| Salvia L. | 1 / 4 / 900 | |
| Satureja L. | 0 / 1 / 30 | |
| Scutellaria L. | 0 / 2 / 300 | |
| Solenostemon Thonn. (Plectranthus) | 1 / 2 / 60 | nat. cult. orn. |
| Stachys L. | 0 / 1 / 300 | |
| Teucrium L. | 0 / 3 / 100 | |

## LAURACEAE  16(278) / 45(2200)

| | | |
|---|---|---|
| Actinodaphne Nees | 5 / 4 / 70 | |
| Alseodaphne Nees | 0 / 1 / 50 | |
| Beilschmiedia Nees | 3 / 15 / 200 | |
| Brassiodendron = Endiandra | | |
| Bryantea = Neolitsea | | |
| Caryodaphnopsis Airy Shaw | 0 / 1 / 7 | |
| Cassytha L. | 1 / 2 / 16 | |
| Cinnamomum Schaeffer | 2 / 21 / 250 | |
| Cryptocarya R. Br. | 18 / 96 / 200 | |
| Dehaasia Blume | 0 / 1 / 35 | |
| Endiandra R. Br. | 9 / 59 / 80 | |
| Lindera Thunb. | 0 / 1 / 80 | |
| Litsea Lam. | 25 / 65 / 400 | |
| Massoia = Cryptocarya | | |
| Neolitsea (Benth.) Merr. | 0 / 6 / 60 | |
| Nothaphoebe Blume | 1 / 2 / 30 | |
| Persea Miller | 1 / 1 / 150 | intro. cult. ed. |
| Phoebe Nees | 0 / 2 / 70 | |
| Pseudocryptocarya = Cryptocarya | | |
| Tetradenia = Neolitsea | | |
| Tetranthera = Litsea | | |
| Triadodaphne Kosterm. | 1 / 1 / 1 | |

## Lecythidaceae see Barringtoniaceae

## LEEACEAE  1(18) / 1(34)

| | |
|---|---|
| Leea Royen ex L. | 6 / 18 / 34 |

## Leguminosae see Caesalpiniaceae, Fabaceae, Mimosaceae

## LEMNACEAE  3(4) / 6(30)

| | |
|---|---|
| Lemna L. | 0 / 2 / 9 |
| Spirodela Schleiden | 0 / 1 / 4 |
| Wolffia Horkel ex Schleiden | 0 / 1 / 7 |

## LENTIBULARIACEAE  1(6) / 4(245)

| | |
|---|---|
| Utricularia L. | 0 / 6 / 180 |

## LILIACEAE    13(32) / 294(4550)

(see also Agavaceae, Dracaenaceae, Sansevieraceae, Smilacaceae, Xanthorrhoeaceae)

| | | |
|---|---|---|
| **Allium** L. | 2 / 2 / 700 | intro. cult. ed. |
| Aphoma = Iphigenia | | |
| **Arthropodium** R. Br. | 0 / 1 / 13 | |
| **Asparagus** L. | 1 / 1 / 100 | intro. cult. ed. |
| **Astelia** Banks & Sol. ex R. Br. | 0 / 2 / 25 | |
| **Caesia** R. Br. | 0 / 1 / 13 | |
| Chlorophytum Ker-Gawler | no record | |
| **Dianella** Lam. | 1 / 17 / 25 | |
| Dichopogon = Arthropodium | | |
| **Gloriosa** L. | 1 / 1 / 1 | nat. cult. orn. |
| Heckelia = Rhipogonum | | |
| **Iphigenia** Kunth | 0 / 1 / 9 | |
| **Ripogonum** Forster & Forster f. (Rhipogonum) | 0 / 2 / 8 | |
| **Romnalda** Harvey | 0 / 1 / 2 | |
| **Schelhammera** R. Br. | 0 / 1 / 3 | |
| **Thysanotus** R. Br. | 0 / 1 / 47 | |
| **Tricoryne** R. Br. | 0 / 1 / 6 | |

## LIMNOCHARITACEAE    1(1) / 3(12)

| | |
|---|---|
| **Butomopsis** Kunth (Tenagocharis) | 0 / 1 / 1 |
| Tenagocharis = Butomopsis | |

## LINACEAE    3(6) / 15(300)

(see also Ixonanthaceae)

| | |
|---|---|
| **Ctenolophon** Oliver | 0 / 1 / 3 |
| Discogyne = Ixonanthes (Ixonanthaceae) | |
| **Durandea** Planchon (Hugonia) | 3 / 4 / 15 |
| **Hugonia** L. | 1 / 1 / 32 |

## LOBELIACEAE    1(10) / 5(950)

| | |
|---|---|
| **Lobelia** L. | 1 / 10 / 365 |
| Pratia = Lobelia | |

## LOGANIACEAE    8(77) / 29(600)

| | |
|---|---|
| Buddleia = Buddleja | |
| **Buddleja** L. | 0 / 2 / 100 |
| Couthovia = Neuburgia | |
| Crateriphytum = Neuburgia | |
| Cynoctonum = Mitreola | |
| **Fagraea** Thunb. | 6 / 30 / 35 |
| **Geniostoma** Forster & Forster f. | 3 / 23 / 52 |
| **Mitrasacme** Labill. | 0 / 7 / 40 |

| | |
|---|---|
| Mitreola L. | 0 / 3 / 6 |
| Neuburgia Blume | 2 / 6 / 12 |
| Spigelia L. | 0 / 1 / 50 |
| Strychnos L. | 3 / 5 / 190 |

Lophopyxaceae see Celastraceae

## LORANTHACEAE 15(58) / 70(940)

(see also Viscaceae)

| | | |
|---|---|---|
| Amyema Tieghem | 2 / 25 / 90 | |
| Amylotheca Tieghem | 4 / 2 / 5 | |
| Cecarria Barlow | 0 / 1 / 1 | endemic in New Guinea |
| Dactyliophora Tieghem (Dactylophora) | 3 / 3 / 3 | |
| Decaisnina Tieghem | 1 / 9 / 30 | |
| Dendrophtoe C. Martius | 1 / 4 / 30 | |
| Dicymanthes = Amyema | | |
| Distrianthes Danser | 1 / 1 / 1 | endemic in New Guinea |
| Elytranthe see Decaisnina, Macrosolen | | |
| Lepeostegeres Blume | 0 / 1 / 13 | |
| Loranthus Jacq. | 0 / 1 / 1 | |
| Macrosolen (Blume) Blume | 0 / 1 / 25 | |
| Muellerina see Cecarria | | |
| Notanthera (DC.) G. Don f. (Loranthus) | 0 / 1 / 1 | |
| Papuanthes Danser | 0 / 1 / 1 | endemic in New Guinea |
| Phrygilanthus = Notanthera | | |
| Rhizomonanthes Danser | 0 / 3 / 3 | endemic in New Guinea |
| Sogerianthe Danser | 3 / 4 / 4 | endemic in Papuasia |
| Tetradyas Danser | 0 / 1 / 1 | endemic in Papuasia |
| Ungula = Amyema | | |

## LYTHRACEAE 6(20) / 26(580)

| | | |
|---|---|---|
| Ammannia L. | 1 / 5 / 30 | |
| Lagerstroemia L. | 1 / 9 / 53 | intro. cult. orn. |
| Lawsonia L. | 0 / 1 / 1 | |
| Lythrum L. | 0 / 1 / 38 | |
| Pemphis Forster & Forster f. | 1 / 1 / 2 | |
| Portula = Lythrum | | |
| Rotala L. | 0 / 3 / 44 | |

## MAGNOLIACEAE 3(4) / 7(200)

| | | |
|---|---|---|
| Aromadendron = Magnolia | | |
| Elmerrillia Dandy | 0 / 2 / 4 | |
| Magnolia L. | 0 / 1 / 125 | |
| Michelia L. | 0 / 1 / 45 | intro. cult. |
| Talauma = Magnolia | | |

## MALPIGHIACEAE
3(7) / 68(1100)

| | | |
|---|---|---|
| Heteropteris see Rhyssopteris | | |
| Malpighia L. | 1 / 1 / 40 | intro. cult. orn. |
| Rhyssopteris Blume ex A. Juss. (Rhyssopterys) | 0 / 5 / 6 | |
| Ryssopterys = Rhyssopteris | | |
| Tristellateia Thouars | 1 / 1 / 20 | intro. cult. orn. |

## MALVACEAE
9(56) / 121(1550)

| | | |
|---|---|---|
| Abelmoschus Medik. (Hibiscus) | 1 / 5 / 15 | intro. cult. ed. |
| Abutilon Miller | 0 / 4 / 100 | |
| Cephalohibiscus = Thespesia | | |
| Gossypium L. | 0 / 2 / 39 | cult. econ. |
| Hibiscus L. | 7 / 24 / 200 | 3 intro. cult. orn. |
| Malachra L. | 0 / 1 / 9 | |
| Malvastrum A. Gray | 1 / 1 / 14 | |
| Sida L. | 2 / 8 / 150 | nat. |
| Thespesia Sol. ex Corr. Serr. | 3 / 8 / 17 | |
| Urena L. | 2 / 2 / 6 | |
| Wilhelminia = Hibiscus | | |

## MARANTACEAE
7(18) / 31(350)

(see also Heliconiaceae)

| | | |
|---|---|---|
| Actoplanes = Donax | | |
| Clinogyne see Donax, Marantochloa, Schumannianthus | | |
| Cominsia Hemsley | 2 / 4 / 5 | |
| Donax Lour. | 1 / 2 / 6 | |
| Maranta L. | 1 / 1 / 20 | nat. cult. |
| Marantochloa Brongn. ex Gris | 0 / 1 / 15 | |
| Phacelophrynium Schumann | 0 / 2 / 9 | |
| Phrynium Willd. | 0 / 7 / 20 | |
| Schumannianthus Gagnepain | 0 / 1 / 2 | |
| Thalia see Donax | | |

Mastixiaceae see Cornaceae

## MELASTOMATACEAE
22(235) / 215(4750)

| | | |
|---|---|---|
| Allomorphia = Oxyspora | | |
| Anplectrum = Diplectria | | |
| Astronia Blume | 0 / 38 / 70 | |
| Astronidium A. Gray | 14 / 30 / 35 | |
| Bamlera = Astronidium | | |
| Beccarianthus see Astronidium | | |
| Catanthera F. Muell. | 0 / 9 / 10 | |
| Clidemia D. Don | 1 / 1 / 165 | nat. |
| Creochiton Blume | 0 / 2 / 6 | |

| | | |
|---|---|---|
| **Dicerospermum** Bakh. f. | 1 / 1 / 1 | endemic in New Guinea |
| **Diplectria** (Blume) Reichb. | 0 / 2 / 11 | |
| **Dissochaeta** Blume | 0 / 3 / 20 | |
| Erpetina = Medinilla | | |
| Everettia see Astronidium | | |
| Hederella = Catanthera | | |
| Hypenanthe (Blume) Blume | no record | |
| Kibessia = Pternandra | | |
| **Macrolenes** Naudin ex. Miq. | 0 / 1 / 15 | |
| Marumia = Macrolenes | | |
| **Medinilla** Gaudich. | 20 / 84 / 150 | |
| **Melastoma** L. | 4 / 10 / 70 | |
| **Memecylon** L. | 1 / 9 / 150 | |
| **Neodissochaeta** Bakh. f. (Dissochaeta) | 0 / 2 / 10 | |
| **Ochtocharis** Blume | 1 / 3 / 7 | |
| Omphalopus see Dissochaeta | | |
| **Osbeckia** L. | 0 / 3 / 60 | |
| **Otanthera** Blume | 0 / 8 / 15 | |
| **Oxyspora** DC. | 0 / 6 / 24 | |
| **Pachycentria** Blume | 0 / 1 / 8 | |
| Phyllapophysis = Catanthera | | |
| **Pogonanthera** Blume | 0 / 2 / 4 | |
| **Poikilogyne** Baker f. | 1 / 11 / 20 | |
| **Pternandra** Jack | 0 / 4 / 15 | |
| Scrobicularia = Poikilogyne | | |
| **Sonerila** Roxb. | 0 / 1 / 100 | |

## MELIACEAE

| | | |
|---|---|---|
| | 19(174) / 51(575) | |
| **Aglaia** Lour. | 12 / 66 / 100 | |
| Aglaiopsis = Aglaia | | |
| Amoora = Aglaia | | |
| **Anthocarapa** Pierre | 1 / 1 / 1 | |
| **Aphanamixis** Blume | 5 / 1 / 3 | |
| **Azadirachta** A. Juss. | 0 / 2 / 2 | |
| **Carapa** Aublet (Xylocarpus) | 0 / 1 / 2 | |
| **Cedrela** P. Browne (Toona) | 2 / 2 / 8 | intro. cult. |
| **Chisocheton** Blume | 4 / 27 / 51 | |
| Dasycoleum = Chisocheton | | |
| Didymocheton = Dysoxylum | | |
| **Dysoxylum** Blume | 24 / 54 / 75 | |
| Epicharis = Dysoxylum | | |
| Goniocheton = Dysoxylum | | |
| Hearnia = Aglaia | | |
| **Khaya** A. Juss. | 1 / 1 / 7 | intro. cult. |

| | | |
|---|---|---|
| **Lansium** Corr. Serr. | 0 / 1 / 3 | |
| **Melia** L. | 2 / 2 / 3 | intro. |
| Melio-Schinzia = Chisocheton | | |
| Nymania see Phyllanthus (Euphorbiaceae) | | |
| **Pseudocarapa** Hemsley (Dysolxylum) | 1 / 1 / 1 | |
| **Reinwardtiodendron** Koord. | 0 / 1 / 7 | |
| **Sandoricum** Cav. | 0 / 1 / 5 | |
| **Swietenia** Jacq. | 1 / 1 / 3 | intro. cult. |
| Synoum A. Juss. | no record | |
| **Toona** (Endl.) M. Roemer | 2 / 2 / 6 | 1 intro. cult. |
| Trichilia P. Browne | no record | |
| **Turraea** L. | 0 / 3 / 60 | |
| **Vavaea** Benth. | 4 / 4 / 4 | |
| Walsura see Trichilia | | |
| **Xylocarpus** Koenig | 3 / 3 / 3 | |

Meliosmaceae see Sabiaceae

Memecylaceae see Melastomataceae

# MENISPERMACEAE

| | | |
|---|---|---|
| **MENISPERMACEAE** | 19(33) / 78(525) | |
| **Albertisia** Becc. | 0 / 1 / 17 | |
| **Anamirta** Colebr. | 0 / 1 / 3 | |
| **Arcangelisia** Becc. | 0 / 2 / 2 | |
| Bania = Carronia | | |
| **Carronia** F. Muell. | 0 / 1 / 4 | |
| **Chlaenandra** Miq. | 0 / 1 / 1 | endemic in New Guinea |
| **Cissampelos** L. | 0 / 1 / 19 | |
| **Diploclisia** Miers | 0 / 1 / 4 | |
| **Hypserpa** Miers | 1 / 2 / 20 | |
| **Legnephora** Miers | 1 / 3 / 5 | |
| **Limacia** Lour. | 0 / 1 / 3 | ident.? |
| **Macrococculus** Becc. | 0 / 1 / 1 | endemic in New Guinea |
| **Pachygone** Miers | 0 / 1 / 12 | |
| **Parabaena** Miers | 2 / 1 / 6 | |
| **Pericampylus** Miers | 0 / 1 / 7 | |
| Porotheca = Chlaenandra | | |
| **Pycnarrhena** Miers ex Hook. f. & Thomson | 1 / 3 / 9 | |
| **Sarcopetalum** F. Muell. | 0 / 1 / 1 | |
| **Stephania** Lour. | 3 / 6 / 40 | |
| **Tinomiscium** Miers ex Hook. f. | 0 / 1 / 8 | |
| **Tinospora** Miers | 1 / 4 / 32 | |

# MENYANTHACEAE

| | |
|---|---|
| **MENYANTHACEAE** | 2(6) / 5(40) |
| **Nymphoides** Hill | 0 / 5 / 20 |
| **Villarsia** Vent. | 0 / 1 / 17 |

## MIMOSACEAE                                     19(116) / 58(3100)

| | | |
|---|---|---|
| Abarema = Archidendron | | |
| **Acacia** Miller | 7 / 14 / 1200 | some intro. cult. |
| **Adenanthera** L. | 1 / 3 / 4 | |
| **Albizia** Durazz. | 5 / 17 / 150 | some intro. cult. |
| **Archidendron** F. Muell. | 6 / 45 / 130 | |
| **Calliandra** Benth. | 2 / 3 / 200 | intro. cult. orn. |
| **Cathormion** (Benth.) Hassk. | 0 / 1 / 12 | |
| Cylindrokelupha = Archidendron | | |
| **Desmanthus** Willd. (Neptunia) | 1 / 1 / 25 | intro. |
| **Entada** Adans. | 2 / 2 / 30 | |
| **Enterolobium** C. Martius | 0 / 1 / 5 | intro. cult. |
| Hansemannia = Archidendron | | |
| Inga = Cathormion | | |
| **Leucaena** Benth. | 2 / 8 / 40 | nat. cult. |
| **Mimosa** L. | 3 / 4 / 400 | nat. cult. orn. |
| Morolobium = Archidendron | | |
| **Neptunia** Lour. | 0 / 1 / 12 | |
| Ortholobium = Archidendron | | |
| **Parkia** R. Br. | 0 / 1 / 40 | |
| Piptadenia = Schleinitzia | | |
| **Pithecellobium** C. Martius | 0 / 1 / 20 | intro. cult. |
| **Prosopis** L. | 0 / 2 / 44 | |
| **Samanea** (Benth.) Merr. | 1 / 1 / 7 | intro. cult. |
| **Schleinitzia** Warb. ex Harms (Prosopis) | 1 / 1 / 3 | |
| Serialbizzia = Albizia | | |
| **Serianthes** Benth. | 4 / 4 / 10 | |
| **Xylia** Benth. (Esclerona) | 1 / 1 / 13 | intro. cult. |

## MITRASTEMMATACEAE                          1(1) / 1(2)

| | | |
|---|---|---|
| **Mitrastemma** Makino (Mitrastemmon) | 0 / 1 / 2 | |

## MOLLUGINACEAE                               2(3) / 9(90)

| | | |
|---|---|---|
| **Glinus** L. | 0 / 1 / 6 | |
| **Mollugo** L. | 1 / 2 / 15 | |

## MONIMIACEAE                                 11(89) / 35(450)

| | | |
|---|---|---|
| Anthobembix = Steganthera | | |
| Daphnandra see Dryadodaphne | | |
| **Dryadodaphne** S. Moore | 0 / 2 / 2 | |
| **Faikea** Philipson (Faika) | 0 / 1 / 1 | endemic in New Guinea |
| **Hedycarya** Forster & Forster f. | 1 / 3 / 13 | |
| Isomerocarpa = Dryadodaphne | | |
| **Kairoa** Philipson | 0 / 1 / 1 | endemic in New Guinea |
| **Kibara** Endl. | 0 / 37 / 60 | |

| | | |
|---|---|---|
| **Lauterbachia** Perkins | 0 / 1 / 1 | endemic in New Guinea |
| **Levieria** Becc. | 0 / 7 / 18 | |
| Matthaea Blume | no record | |
| **Mollinedia** Ruíz & Pavón (Wilkiea) | 0 / 1 / 95 | |
| **Palmeria** F. Muell. | 0 / 11 / 15 | |
| **Steganthera** Perkins | 2 / 16 / 24 | |
| **Wilkiea** F. Muell. | 0 / 1 / 6 | |

## MORACEAE

| | | |
|---|---|---|
| | 15(165) / 48(1200) | |
| **Antiaris** Leschen. | 2 / 2 / 4 | |
| **Antiaropsis** Schumann | 0 / 1 / 1 | endemic in New Guinea |
| **Artocarpus** Foster & Foster f. | 4 / 5 / 31 | 1 intro. cult. |
| **Broussonetia** L'Hérit. ex Vent. | 1 / 1 / 7 | intro. |
| Calpidochlamys see Trophis | | |
| **Castilla** Sessé | 0 / 1 / 3 | intro. |
| Castilloa = Castilla | | |
| Caturus = Malaisia | | |
| Chlorophora = Milicia | | |
| **Cudrania** Trécul (Maclura) | 0 / 1 / 5 | |
| Dammaropsis = Ficus | | |
| **Fatoua** Gaudich. | 1 / 1 / 1 | |
| **Ficus** L. | 73 / 138 / 800 | some cult. |
| **Maclura** Nutt. | 0 / 2 / 2 | |
| **Malaisia** Blanco | 0 / 1 / 1 | |
| **Milicia** Sim | 1 / 1 / 5 | intro. cult. |
| Morus see Streblus | | |
| **Parartocarpus** Baillon | 1 / 2 / 4 | |
| Paratrophis = Streblus | | |
| **Prainea** King ex Hook. f. | 1 / 1 / 4 | |
| Pseudomorus = Streblus | | |
| Pseudotrophis = Streblus | | |
| **Streblus** Lour. | 2 / 5 / 22 | 1 intro. cult. ed. |
| **Trophis** P. Browne | 0 / 3 / 11 | |

## MORINGACEAE

| | | |
|---|---|---|
| | 1(1) / 1(10) | |
| **Moringa** Adans. | 1 / 1 / 10 | nat. |

## MUSACEAE

| | | |
|---|---|---|
| | 2(15) / 2(42) | |

(see also Heliconiaceae)

| | | |
|---|---|---|
| **Ensete** Horan. | 0 / 1 / 7 | |
| **Musa** L. | 7 / 14 / 35 | |

## MYOPORACEAE

| | | |
|---|---|---|
| | 1(1) / 5(220) | |
| **Myoporum** Banks & Sol. ex Forster f. | 0 / 1 / 32 | |

## MYRICACEAE                     1(1) / 3(50)
Myrica L.                         0 / 1 / 50

## MYRISTICACEAE                  5(79) / 19(40)
Endocomia Wilde (Horsfieldia)     0 / 1 / 4
Gymnacranthera Warb.             0 / 1 / 17
Horsfieldia Willd.               8 / 37 / 80
Knema Lour.                      0 / 1 / 83
Myristica Gronov.                19 / 39 / 80

## MYRSINACEAE                    14(179) / 39(1250)
Abromeitia = Fittingia
Aegiceras Gaertner               1 / 4 / 2
Amblyanthus A. DC. (Conandrium)  0 / 1 / 4
Ardisia Sw.                      5 / 42 / 250
Conandrium (Schumann) Mez        0 / 9 / 9
Discocalyx (A. DC.) Mez          3 / 15 / 50
Embelia Burm. f.                 0 / 12 / 130
Fittingia Mez                    0 / 5 / 5          endemic in New Guinea
Grenacheria Mez                  0 / 2 / 10
Labisia Lindley                  0 / 1 / 9
Loheria Merr.                    0 / 2 / 6
Maesa Forssk.                    5 / 46 / 100
Myrsine L. (Rapanea)             1 / 1 / 5
Pimelandra = Ardisia
Rapanea Aublet (Myrsine)         4 / 30 / 150
Tapeinosperma Hook. f.           2 / 8 / 40

## MYRTACEAE                      28(308) / 121(3850)
Acmena = Syzygium
Agonis (DC.) Sweet (Sinoga)      0 / 1 / 12
Aphanomyrtus = Syzygium
Asteromyrtus = Melaleuca
Backhousia Hook. & Harvey        0 / 1 / 8
Baeckea L.                       0 / 1 / 90
Basisperma C. White             0 / 1 / 1          endemic in New Guinea
Calyptranthus = Syzygium
Cleistocalyx = Syzygium
Decaspermum Forster & Forster f. 3 / 15 / 30
Eucalyptopsis C. White           0 / 2 / 2
Eucalyptus L'Hérit.              9 / 9 / 450        intro. cult. econ.
Eugenia L.                       19 / 32 / 1000
Fenzlia = Myrtella
Jambosa = Syzygium
Jossinia = Eugenia

**Kania** Schltr. (Metrosideros) — 1 / 3 / 3
**Kjellbergiodendron** Burret — 0 / 1 / 1
**Leptospermum** Forster & Forster f. — 0 / 3 / 30
**Lindsayomyrtus** B. Hyland & Steenis — 0 / 1 / 1
Mearnsia = Metrosideros
**Melaleuca** L. — 1 / 10 / 150
**Metrosideros** Banks ex Gaertner — 7 / 10 / 50
**Myrtella** F. Muell. — 1 / 2 / 9
**Myrtus** L. — 0 / 13 / 20
Nani see Xanthostemon
Nelitris = Decaspermum
**Octamyrtus** Diels — 0 / 5 / 6
**Osbornia** F. Muell. — 1 / 1 / 1
Pseudoeugenia = Syzygium
**Psidium** L. — 1 / 2 / 100 — cult. ed.
**Rhodamnia** Jack — 4 / 17 / 23
**Rhodomyrtus** (DC.) Reichb. — 2 / 6 / 11
**Sinoga** S.T. Blake — 0 / 1 / 1
**Syncarpia** Ten. (Metrosideros) — 0 / 1 / 5
**Syzygium** Gaertner — 23 / 138 / 500
**Tristania** R. Br. (Tristaniopsis) — 0 / 5 / 8
**Uromyrtus** Burret — 0 / 2 / 15
**Xanthomyrtus** Diels — 1 / 20 / 23
**Xanthostemon** F. Muell. — 1 / 5 / 40
Xenodendron = Syzygium

## NAJADACEAE — 1(5) / 1(35)
**Najas** L. — 0 / 5 / 35

Naucleaceae see Rubiaceae

## NELUMBONACEAE — 1(1) / 1(2)
**Nelumbo** Adans. (Nelumbium) — 0 / 1 / 2

## NEPENTHACEAE — 1(11) / 1(70)
**Nepenthes** L. — 0 / 11 / 70

## NYCTAGINACEAE — 5(27) / 34(350)
**Boerhavia** L. — 3 / 3 / 40
**Bougainvillea** Comm. ex Juss. — 2 / 4 / 14 — intro. cult. orn.
Calpidia see Ceodes
**Ceodes** Forster & Forster f. — 1 / 5 / 25
Heimerliodendron Skottsb. = Pisonia
**Mirabilis** L. — 1 / 1 / 45 — intro. cult. orn.
**Pisonia** L. — 4 / 14 / 35

## NYMPHAEACEAE                    2(7) / 6(60)

(see also Nelumbonaceae)

Barclaya = Hydrostemma

**Hydrostemma** Wallich              0 / 1 / 2

**Nymphaea** L.                      0 / 6 / 35

## NYSSACEAE                        1(1) / 3(8)

**Nyssa** L.                         0 / 1 / 5

## OCHNACEAE                        2(3) / 37(460)

**Brackenridgea** A. Gray            0 / 1 / 7

**Schuurmansia** Blume               1 / 2 / 3

## OLACACEAE                        4(5) / 29(200)

(see also Opiliaceae)

**Anacolosa** (Blume) Blume          1. / 2 / 16

**Erythropalum** Blume               0 / 1 / 1

Harmandia Baillon                    no record

**Olax** L.                          1 / 1 / 40

Strombosia Blume                     no record

Villaresia see Gonocaryum (Icacinaceae)

**Ximenia** L.                       1 / 1 / 8

## OLEACEAE                         5(32) / 24(900)

**Chionanthus** L. (Linociera)       9 / 10 / 120

**Jasminum** L.                      4 / 13 / 450

**Ligustrum** L.                     0 / 3 / 50

Linociera = Chionanthus

**Myxopyrum** Blume                  0 / 4 / 4

**Olea** L.                          0 / 2 / 20

Phillyrea see Chionanthus

Visiania = Ligustrum

## ONAGRACEAE                       2(7) / 24(650)

**Epilobium** L.                     0 / 4 / 200

Jussiaea = Ludwigia

**Ludwigia** L.                      1 / 3 / 75

## OPILIACEAE                       5(7) / 9(28)

**Cansjera** Juss.                   1 / 2 / 3

**Champereia** Griffith              0 / 1 / 1

**Gjellerupia** Lauterb.             0 / 1 / 1        endemic in New Guinea

**Lepionurus** Blume                 0 / 1 / 1

**Opilia** Roxb.                     0 / 2 / 2

## ORCHIDACEAE    150(2806) / 795(17500)

| | | |
|---|---|---|
| **Acanthophippium** Blume | 0 / 3 / 15 | |
| **Acianthus** R. Br. | 1 / 2 / 20 | |
| **Acriopsis** Blume | 2 / 2 / 12 | |
| **Adenoncos** Blume | 0 / 3 / 10 | |
| **Aerides** Lour. | 0 / 2 / 40 | |
| **Aglossorhyncha** Schltr. | 1 / 10 / 13 | |
| **Agrostophyllum** Blume | 5 / 49 / 60 | |
| **Angraecopsis** Kraenzlin | 1 / 1 / 16 | |
| **Angraecum** Bory | 1 / 1 / 206 | |
| **Anoectochilus** Blume | 0 / 4 / 25 | |
| **Aphyllorchis** Blume | 0 / 4 / 20 | |
| Aporum see Dendrobium | | |
| **Apostasia** Blume | 0 / 1 / 11 | |
| **Appendicula** Blume | 10 / 48 / 100 | |
| **Arachnis** Blume | 0 / 2 / 7 | |
| **Arthrochilus** F. Muell. (Spiculaea) | 0 / 1 / 4 | |
| Ascocentrum Schltr. ex J.J. Sm. | no record | |
| **Ascoglossum** Schltr. (Sarcochilus) | 1 / 2 / 2 | endemic in Papuasia |
| Aulostylis = Calanthe | | |
| **Bletia** Ruíz & Pavón | 0 / 1 / 45 | |
| **Bogoria** J.J. Sm. | 0 / 2 / 3 | |
| **Bromheadia** Lindley | 0 / 2 / 11 | |
| **Bulbophyllum** Thouars | 17 / 500 / 1200 | |
| **Cadetia** Gaudich. | 6 / 33 / 55 | endemic in Papuasia |
| **Calanthe** R. Br. | 9 / 51 / 120 | |
| Callista = Dendrobium | | |
| **Calochilus** R. Br. | 0 / 1 / 5 | |
| **Calymmanthera** Schltr. | 1 / 3 / 3 | endemic in New Guinea |
| Camarotis = Micropera | | |
| Carteretia = Cleisostoma | | |
| **Centrostigma** Schltr. | 1 / 1 / 3 | |
| **Ceratochilus** Blume | 0 / 1 / 1 | ident.? |
| **Ceratostylis** Blume | 3 / 56 / 60 | |
| Cestichis = Liparis | | |
| **Chamaeanthus** Schltr. ex J.J. Sm. | 1 / 4 / 14 | |
| **Cheirostylis** Blume | 0 / 2 / 22 | |
| Chilopogon = Appendicula | | |
| **Chitonanthera** Schltr. | 0 / 11 / 11 | endemic in New Guinea |
| **Chitonochilus** Schltr. | 0 / 1 / 1 | endemic in New Guinea |
| **Chrysoglossum** Blume | 1 / 3 / 4 | |
| Cirrhopetalum = Bulbophyllum | | |
| **Claderia** Hook. f. | 0 / 1 / 2 | |
| **Cleisostoma** Blume (Sarcochilus) | 2 / 9 / 100 | |

| | | |
|---|---|---|
| **Codonosiphon** Schltr. | 0 / 3 / 3 | endemic in New Guinea |
| **Coelogyne** Lindley | 6 / 16 / 200 | |
| **Collabium** Blume | 0 / 1 / 11 | |
| **Corybas** Salisb. | 3 / 38 / 50 | |
| Corymbis = Corymborkis | | |
| **Corymborkis** Thouars | 1 / 3 / 32 | |
| Corysanthes = Corybas | | |
| **Cryptostylis** R. Br. | 1 / 9 / 15 | |
| **Cymbidium** Sw. | 1 / 4 / 40 | |
| **Cynorkis** Thouars | 2 / 2 / 125 | |
| **Cyphochilus** Schltr. | 0 / 7 / 7 | endemic in New Guinea |
| Cypripedium see Paphiopedilum | | |
| Cyrtopera = Eulophia | | |
| **Cyrtorchis** Schltr. | 1 / 1 / 15 | |
| **Cyrtosia** Bl. (Galeola) | 0 / 1 / 1 | |
| Cystopus = Pristiglottis | | |
| **Cystorchis** Blume | 0 / 6 / 21 | |
| **Dactylorhynchus** Schltr. | 0 / 1 / 1 | endemic in New Guinea |
| **Dendrobium** Sw. | 70 / 500 / 900 | |
| **Dendrochilum** Blume | 0 / 5 / 120 | |
| Dendrocolla = Sarcochilus | | |
| Desmotrichum = Flickingeria | | |
| Dichopus = Dendrobium | | |
| **Didymoplexis** Griffith | 0 / 3 / 10 | |
| Dienia = Malaxis | | |
| Diglyphosa Blume | no record | |
| **Diplocaulobium** (Reichb. f.) Kraenzlin | | |
| (Dendrobium) | 5 / 61 / 70 | |
| **Dipodium** R. Br. | 3 / 3 / 22 | |
| **Disperis** Sw. | 1 / 1 / 75 | |
| **Doritis** Lindley | 0 / 1 / 2 | |
| **Dossinia** Morren | 0 / 1 / 1 | |
| **Drakaea** Lindley | 0 / 1 / 4 | |
| **Dryadorchis** Schltr. | 0 / 2 / 2 | endemic in New Guinea |
| **Earina** Lindley | 0 / 2 / 7 | |
| Ephemerantha = Flickingeria | | |
| Ephippium = Bulbophyllum | | |
| **Epiblastus** Schltr. | 1 / 12 / 15 | |
| **Epidendrum** L. | 0 / 2 / 500 | |
| **Epigeneium** Gagnepain (Dendrobium) | 0 / 2 / 35 | |
| **Epipactis** Zinn (Goodyera) | 0 / 1 / 24 | |
| **Epipogium** J. Gmelin ex Borkh. | 0 / 2 / 3 | |
| **Eria** Lindley | 8 / 61 / 350 | |
| **Erythrodes** Blume | 0 / 15 / 60 | |

| | | |
|---|---|---|
| **Eucosia** Blume | 0 / 1 / 2 | |
| **Eulophia** R. Br. | 3 / 13 / 200 | |
| **Eurycentrum** Schltr. | 1 / 5 / 5 | endemic in Papuasia |
| **Flickingeria** A. Hawkes | 4 / 18 / 70 | |
| **Galeola** Lour. | 2 / 4 / 25 | |
| **Gastrodia** R. Br. | 0 / 1 / 20 | |
| Gastroglottis = Liparis | | |
| Geissanthera = Microtatorchis | | |
| **Geodorum** Jackson | 1 / 2 / 16 | |
| Giulianettia = Glossorhyncha | | |
| **Glomera** Blume | 2 / 45 / 48 | |
| **Glossorhyncha** Ridley (Glomera) | 1 / 44 / 80 | |
| **Goodyera** R. Br. | 4 / 22 / 40 | |
| **Grammatophyllum** Blume | 4 / 8 / 10 | |
| **Gynoglottis** J.J. Sm. | 0 / 1 / 1 | |
| **Habenaria** Willd. | 8 / 38 / 500 | |
| Haplochilus = Bulbophyllum | | |
| **Hetaeria** Blume | 7 / 9 / 20 | |
| **Hippeophyllum** Schltr. | 0 / 2 / 5 | |
| Hologyne = Coelogyne | | |
| Hyalosema = Bulbophyllum | | |
| **Hylophila** Lindley | 1 / 2 / 4 | |
| **Hymenorchis** Schltr. | 0 / 6 / 6 | endemic in New Guinea |
| **Ischnocentrum** Schltr. | 0 / 1 / 1 | endemic in New Guinea |
| Katharinea = Epigeneium | | |
| **Kuhlhasseltia** J.J. Sm. | 0 / 2 / 5 | |
| Latouria = Dendrobium | | |
| **Lecanorchis** Blume | 0 / 5 / 20 | |
| Lectandra = Poaephyllum | | |
| **Lepidogyne** Blume | 0 / 3 / 3 | |
| Linodorum = Epipactis | | |
| **Liparis** Rich. | 4 / 94 / 250 | |
| Lobogyne = Appendicula | | |
| **Luisia** Gaudich. | 1 / 3 / 30 | |
| **Macodes** (Blume) Lindley | 4 / 6 / 10 | |
| **Malaxis** Sol. ex Sw. | 7 / 84 / 300 | |
| **Malleola** J.J. Sm. & Schltr. | 1 / 5 / 20 | |
| Maxillaria Ruíz & Pavón | no record | |
| **Mediocalcar** J.J. Sm. | 3 / 41 / 60 | |
| **Micropera** Lindley | 1 / 2 / 13 | |
| Microstylis = Malaxis | | |
| **Microtatorchis** Schltr. | 0 / 13 / 20 | |
| Microtis R. Br. | no record | |
| **Miltonia** Lindley | 1 / 1 / 20 | intro.? |

Mischobulbum = Tainia

Moerenhoutia = Malaxis

Monosepalum = Bulbophyllum

Nephelaphyllum see Collabium

**Nervilia** Comm. ex Gaudich.     1 / 13 / 80

**Neuwiedia** Blume     1 / 3 / 7

**Oberonia** Lindley     5 / 95 / 330

**Octarrhena** Thwaites     2 / 22 / 35

Onychium Blume = Dendrobium

Orchiodes = Goodyera

**Ornithochilus** (Lindley) Wallich ex Benth.     0 / 1 / 1

Orsidice = Thrixspermum

Osyricera = Bulbophyllum

Oxyanthera = Thelasis

Oxystophyllum = Dendrobium

Pachyne = Phaius

**Pachystoma** Blume     0 / 3 / 11

**Paphiopedilum** Pfitzer     2 / 7 / 36

**Papuaea** Schltr.     0 / 1 / 1     endemic in New Guinea

**Pedilochilus** Schltr.     1 / 16 / 18     endemic in Papuasia

Pelma = Bulbophyllum

**Peristylus** Blume (Habenaria, Herminium)     2 / 6 / 70

**Phaius** Lour.     1 / 12 / 30

**Phalaenopsis** Blume     1 / 3 / 40

**Pholidota** Lindley ex Hook.     1 / 5 / 29

**Phreatia** Lindley     11 / 108 / 190

Phyllorchis = Bulbophyllum

Physurus = Erythrodes

**Pilophyllum** Schltr. (Chrysoglossum)     1 / 1 / 1

**Platanthera** Rich. (Habenaria)     0 / 2 / 85

Platyclinis see Dendrochilum

**Platycoryne** Riechb. f.     2 / 2 / 18

**Platylepis** A. Rich.     0 / 4 / 10

**Pleione** D. Don (Coelogyne)     0 / 1 / 7

**Plocoglottis** Blume     3 / 26 / 30

**Poaephyllum** Ridley     0 / 4 / 4

**Podochilus** Blume     1 / 34 / 60

**Pogonia** Juss. (Nervilia)     0 / 4 / 10

**Pomatocalpa** Breda, Kuhl & Hasselt     2 / 6 / 60

**Porphyrodesme** Schltr.     0 / 1 / 1     endemic in New Guinea

**Pristiglottis** Cretz. & J.J. Sm.     2 / 4 / 13

**Pseuderia** Schltr.     2 / 11 / 20

Pseudoliparis see Malaxis

Pseudomacodes = Macodes

| | | |
|---|---|---|
| **Pteroceras** Hasselt ex Hassk. (Sarcochilus) | 1 / 3 / 30 | |
| **Pterostylis** R. Br. | 0 / 3 / 60 | |
| **Rangaeris** (Schltr.) Summerh. | 1 / 1 / 6 | intro. |
| **Renanthera** Lour. | 1 / 2 / 10 | |
| Rhynchophreatia = Phreatia | | |
| **Ridleyella** Schltr. | 0 / 1 / 1 | endemic in New Guinea |
| **Robiquetia** Gaudich. | 3 / 5 / 20 | |
| **Saccoglossum** Schltr. | 0 / 2 / 2 | endemic in New Guinea |
| **Saccolabiopsis** J.J. Sm. | 0 / 5 / 8 | |
| **Saccolabium** Blume | 1 / 28 / 36 | |
| Sarcanthus = Cleisostoma | | |
| **Sarcochilus** R. Br. | 2 / 10 / 12 | |
| Sarcopodium see Dendrobium | | |
| Sayeria see Dendrobium | | |
| **Schoenorchis** Blume | 3 / 3 / 20 | |
| **Sepalosiphon** Schltr. | 0 / 1 / 1 | endemic in New Guinea |
| **Spathoglottis** Blume | 3 / 29 / 40 | |
| Spiculaea Lindley | no record | |
| **Spiranthes** Rich. | 1 / 4 / 42 | |
| Stauropsis = Trichoglottis | | |
| **Stereosandra** Blume | 0 / 1 / 1 | |
| **Taeniophyllum** Blume | 2 / 48 / 90 | |
| **Tainia** Blume | 0 / 3 / 25 | |
| Tapeinoglossum = Bulbophyllum | | |
| **Thelasis** Blume | 1 / 15 / 20 | |
| **Thelymitra** Forster & Forster f. | 1 / 1 / 45 | |
| **Thrixspermum** Lour. | 4 / 14 / 100 | |
| **Trachoma** Garay | 1 / 1 / 6 | |
| **Trichoglottis** Blume | 4 / 6 / 60 | |
| Trichotosia see Eria | | |
| **Tropidia** Lindley | 1 / 9 / 22 | |
| **Vanda** Jones ex R. Br. | 1 / 4 / 35 | |
| **Vandopsis** Pfitzer | 2 / 9 / 10 | |
| **Vanilla** Miller | 1 / 5 / 100 | intro. cult. econ. |
| Vonroemeria = Octarrhena | | |
| **Vrydagzynea** Blume | 5 / 16 / 40 | |
| **Zeuxine** Lindley | 4 / 16 / 77 | |

## OROBANCHACEAE
1(1) / 17(230)

**Aeginetia** L.  0 / 1 / 3

## OXALIDACEAE
3(12) / 8(575)

**Averrhoa** L.  1 / 2 / 2

**Biophytum** DC.  0 / 6 / 50

**Oxalis** L.  1 / 4 / 500

Palmae see Arecaceae

**PANDACEAE**                                   1(1) / 4(18)
Galearia Zoll. & Moritzi                        1 / 1 / 6

**PANDANACEAE**                                 3(114) / 3(675)
Barrotia = Pandanus
Bryanthia = Pandanus
Freycinetia Gaudich.                            22 / 50 / 175
Hombronia = Pandanus
Jeanneretia = Pandanus
Pandanus Parkinson                              36 / 63 / 600
Sararanga Hemsley                               1 / 1 / 2

**PAPAVERACEAE**                                1(1) / 23(210)
Argemone L.                                     0 / 1 / 28                  intro.

Papilionaceae see Fabaceae

**PASSIFLORACEAE**                              3(13) / 18(530)
Adenia Forssk.                                  0 / 2 / 93
Hollrungia Schumann                             1 / 1 / 1
Modecca = Adenia
Passiflora L.                                   5 / 10 / 350        2 intro. cult. ed.
Tacsonia = Passiflora

**PEDALIACEAE**                                 3(3) / 18(95)
Josephinia Vent.                                0 / 1 / 4
Martynia L.                                     0 / 1 / 1                   intro.
Sesamum L.                                      0 / 1 / 15              cult. econ.

**PENTAPHRAGMATACEAE**                          1(1) / 1(25)
Pentaphragma Wallich ex G. Don f.               0 / 1 / 25

Peperomiaceae see Piperaceae

Periplocaceae see Asclepiadaceae

Phellinaceae see Aquifoliaceae

Philesiaceae see Smilacaceae

**PHILYDRACEAE**                                2(2) / 4(5)
Helmholtzia F. Muell.                           0 / 1 / 2
Philydrum Banks ex Gaertner                     0 / 1 / 1

**PHYLLOCLADACEAE**                             1(1) / 1(4)
Phyllocladus Rich. ex Mirbel                    0 / 1 / 4

**PHYTOLACCACEAE**                              2(2) / 18(65)
Phytolacca L.                                   0 / 1 / 25                  intro.

| | | |
|---|---|---|
| Rivina L. | 0 / 1 / 1 | intro. |

## PINACEAE
1(4) / 9(194)

| | | |
|---|---|---|
| Pinus L. | 4 / 4 / 93 | intro. cult. econ. |

## PIPERACEAE
4(126) / 15(2000)

| | | |
|---|---|---|
| Heckeria = Lepianthes | | |
| Lepianthes Raf. (Piper) | 0 / 2 / 10 | |
| Macropiper Miq. (Piper) | 0 / 1 / 9 | |
| Peperomia Ruíz & Pavón | 4 / 23 / 1000 | |
| Piper L. | 18 / 100 / 1000 | some cult. |
| Pothomorphe = Lepianthes | | |

## PITTOSPORACEAE
3(32) / 9(240)

| | | |
|---|---|---|
| Citriobatus A. Cunn. & Putterl. | 0 / 2 / 5 | |
| Hymenosporum R. Br. ex F. Muell. | 0 / 1 / 1 | |
| Pittosporum Banks ex Gaertner | 6 / 29 / 200 | |

## PLANTAGINACEAE
1(7) / 2(255)

| | |
|---|---|
| Plantago L. | 0 / 7 / 250 |

## PLUMBAGINACEAE
2(2) / 22(440)

| | | |
|---|---|---|
| Aegialitis R. Br. | 0 / 1 / 2 | |
| Plumbago L. | 0 / 1 / 10 | cult. orn. |

## POACEAE
132(461) / 737(7950)

| | | |
|---|---|---|
| Acroceras Stapf (Panicum) | no record | |
| Aegopogon Humb. & Bonpl. | 0 / 1 / 3 | |
| Agropyron see Brachypodon | | |
| Agrostis L. | 0 / 6 / 120 | |
| Aira L. | 0 / 1 / 9 | |
| Alloteropsis J. Presl. | 2 / 2 / 5 | nat. |
| Ancistragrostis = Calamagrostis | | |
| Andropogon L. | 0 / 14 / 100 | |
| Aniselytron Merr. | no record | |
| Anthistiria = Themeda | | |
| Anthoxanthum L. | 0 / 2 / 15 | |
| Apluda L. | 1 / 1 / 1 | |
| Aristida L. | 0 / 5 / 260 | |
| Arthraxon Pal. | 0 / 2 / 7 | |
| Arundinaria Michaux | 0 / 1 / 30 | |
| Arundinella Raddi | 0 / 4 / 47 | |
| Arundo L. | 0 / 1 / 3 | cult. orn. |
| Astrebla F. Muell. | no record | |
| Aulacolepis = Aniselytron | | |
| Axonopus see Alloteropsis | | |
| Bambusa Schreber | 3 / 10 / 100 | |

| | | |
|---|---|---|
| **Bothriochloa** Kuntze (Dichanthium) | 0 / 1 / 28 | |
| **Brachiaria** (Trin.) Griseb. | 8 / 8 / 90 | some nat. |
| Brachyachne (Benth.) Stapf | no record | |
| **Brachypodium** Pal. | 0 / 2 / 17 | |
| **Briza** L. | 0 / 1 / 12 | |
| **Bromus** L. | 0 / 2 / 100 | |
| **Buergersiochloa** Pilger | 0 / 1 / 1 | endemic in New Guinea |
| **Calamagrostis** Adans. (Deyeuxia) | 0 / 1 / 250 | |
| **Capillipedium** Stapf (Dichanthium) | 1 / 3 / 14 | |
| **Cenchrus** L. | 3 / 4 / 22 | |
| **Centotheca** Desv. | 2 / 6 / 4 | |
| Chaetochloa = Setaria | | |
| Chamaeraphis see Pseudoraphis | | |
| **Chionachne** R. Br. | 0 / 1 / 5 | |
| **Chionochloa** Zotov | 0 / 1 / 19 | |
| Chloothamnus see Nastus | | |
| **Chloris** Sw. | 2 / 2 / 10 | 1 intro. cult. |
| **Chrysopogon** Trin. | 1 / 2 / 24 | |
| **Cleistochloa** C.E. Hubb. | 0 / 2 / 2 | |
| **Coelachne** R. Br. | 0 / 3 / 5 | |
| **Coelorachis** Brongn. (Mnesithea) | 0 / 2 / 19 | |
| **Coix** L. | 1 / 2 / 6 | |
| Coridochloa = Alloteropsis | | |
| **Cymbopogon** Sprengel | 1 / 3 / 56 | |
| **Cynodon** Rich. | 1 / 2 / 8 | |
| Cynosurus see Dactylocnemium, Eleusine | | |
| **Cyrtococcum** Stapf (Panicum) | 3 / 5 / 11 | |
| **Dactyloctenium** Willd. | 1 / 1 / 13 | |
| **Danthonia** DC. | 0 / 5 / 10 | |
| Dendrocalamus see Bambusa | | |
| **Deschampsia** Pal. | 0 / 2 / 50 | |
| Deyeuxia = Calamagrostis | | |
| **Dichanthium** Willemet | 2 / 3 / 120 | 1 intro. cult. |
| **Dichelachne** Endl. | 0 / 4 / 3 | |
| Digastrium see Ischaemum | | |
| **Digitaria** Haller | 8 / 29 / 170 | |
| **Dimeria** R. Br. | 0 / 7 / 40 | |
| **Dinochloa** Buese | 0 / 1 / 20 | ident.? |
| **Echinochloa** Pal. | 3 / 3 / 20 | |
| **Echinolaena** Desv. (Pseudechinolaena) | 0 / 1 / 6 | |
| **Echinopogon** Pal. | 0 / 1 / 7 | |
| **Ectrosia** R. Br. | 0 / 3 / 12 | |
| Ectrosiopsis see Etrosia | | |
| **Ehrharta** Thunb. | 0 / 2 / 27 | |

| | |
|---|---|
| **Eleusine** Gaertner | 1 / 2 / 9 |
| **Elionurus** Humb. & Bonpl. ex Willd. | 1 / 1 / 14 |
| **Elymus** L. | 0 / 1 / 100 |
| Elyonurus = Elionurus | |
| **Enneapogon** Desv. ex Pal. | 0 / 1 / 30 |
| **Enteropogon** Nees | 0 / 1 / 6 |
| Entolasia see Panicum | |
| **Eragrostis** Wolf | 4 / 12 / 250 |
| **Eremochloa** Buese | 0 / 2 / 10 |
| **Eriachne** R. Br. | 0 / 3 / 35 |
| Erianthus = Saccharum | |
| **Eriochloa** Kunth | 0 / 2 / 30 |
| **Eulalia** Kunth | 0 / 4 / 25 |
| **Festuca** L. | 0 / 6 / 300 |
| **Garnotia** Brongn. | 0 / 8 / 29 |
| **Germainia** Bal. & Poitr. | 0 / 2 / 4 |
| **Gigantochloa** Kurz ex Munro | 0 / 1 / 20 |
| Gymnopogon Pal. | no record |
| Gymnothrix = Pennisetum | |
| **Hackelochloa** Kuntze (Mnesithes) | 1 / 1 / 2 |
| Haplachne = Dimeria | |
| **Hemarthria** R. Br. | 1 / 1 / 12 |
| Hemigymnia see Panicum | |
| **Heteropogon** Pers. | 0 / 3 / 6 |
| Hierochloe see Anthoxanthum | |
| Holcus see Chrysopogon, Sorghum, Andropogon | |
| Homalocenchrus = Leersia | |
| **Hymenachne** Pal. | 0 / 1 / 10 |
| **Hyparrhenia** Andersson ex Fourn. | 0 / 1 / 53 |
| **Ichnanthus** Pal. | 0 / 2 / 25 |
| **Imperata** Cirillo | 1 / 2 / 8 |
| **Isachne** R. Br. | 2 / 15 / 75 |
| **Ischaemum** L. | 4 / 14 / 60 |
| Lagurus see Imperata | |
| **Leersia** Sw. | 0 / 1 / 17 |
| **Leptaspis** R. Br. | 3 / 6 / 7 |
| **Leptochloa** Pal. | 0 / 5 / 20 |
| **Lepturus** R. Br. | 1 / 1 / 12 |
| Lolium L. | no record |
| Lophatherum see Centotheca | |
| Manisuris see Rottboellia | |
| **Melinis** Pal. | 1 / 1 / 11 |
| Microlaena = Ehrharta | |
| **Microstegium** Nees | 1 / 5 / 17 |

Milium see Isachne, Eriochloa

**Miscanthus** Andersson — 1 / 2 / 17

**Monostachya** Merr. (Danthonia) — 0 / 1 / 1

**Muhlenbergia** Schreber (Muehlenbergia) — 0 / 2 / 125

Nardus see Eremochloa

**Nastus** Juss. — 2 / 7 / 9

Neurachne see Sacciolepis

**Ophiuros** Gaertner f. — 0 / 3 / 7

**Oplismenus** Pal. — 2 / 4 / 9

Oreiostachys see Nastus

Ornithocephalochloa = Thuarea

**Oryza** L. — 1 / 7 / 19 — 1 intro. cult. ed.

**Ottochloa** Dandy — 0 / 1 / 6

**Oxytenanthera** Munro — 0 / 1 / 1

**Panicum** L. — 4 / 35 / 600

**Paspalidium** Stapf (Setaria) — 1 / 1 / 27

**Paspalum** L. — 8 / 16 / 250

**Pennisetum** Rich. ex Pers. — 4 / 4 / 70 — 1 intro. cult.

**Perotis** Aiton — 0 / 2 / 10

Phalaris see Arthraxon

**Pheidochloa** S.T. Blake — 0 / 1 / 2

**Phragmites** Adans. — 1 / 3 / 4

**Poa** L. — 0 / 22 / 250

**Pogonatherum** Pal. — 2 / 2 / 2

Pollinia see Eulalia, Microstegium

**Polytoca** R. Br. — 1 / 1 / 6

Polytrias = Eulalia

**Pseudechinolaena** Stapf — 0 / 1 / 6

Pseudopogonatherum = Eulalia

**Pseudoraphis** Griffith — 0 / 2 / 5

**Racemobambos** Holttum — 1 / 6 / 15

Rhaphis = Chrysopogon

**Rhynchelytrum** Nees — 1 / 1 / 15 — nat.

**Rottboellia** L. f. — 1 / 3 / 3 — nat.

**Saccharum** L. — 4 / 5 / 30 — cult. ed.

**Sacciolepis** Nash (Saccolepis) — 0 / 3 / 38

**Schizachyrium** Nees — 0 / 2 / 60

**Schizostachyum** Nees — 2 / 5 / 35

**Sclerandrium** Stapf & C. Hubb. (Germainia) — 0 / 1 / 3

**Sehima** Forssk. — 0 / 1 / 5

**Setaria** Pal. — 4 / 13 / 125 — 1 intro. cult.

**Sorghum** Moench — 2 / 2 / 24 — intro. cult. ed.

Sphaerocaryum Nees ex Hook. f. — no record

**Spinifex** L. — 0 / 1 / 3

| | | |
|---|---|---|
| **Sporobolus** R. Br. | 4 / 7 / 100 | |
| Stegosia = Rottboellia | | |
| **Stenotaphrum** Trin. | 2 / 2 / 7 | |
| Stipa see Spinifex | | |
| Syntherisma = Digitaria | | |
| **Thaumastochloa** C. Hubb. | 0 / 1 / 7 | |
| **Themeda** Forssk. | 2 / 6 / 19 | |
| **Thuarea** Pers. (Thouarea) | 1 / 2 / 2 | |
| **Thysanolaena** Nees | 0 / 1 / 1 | |
| Tricholaena see Rhynchelytrum | | |
| **Tripogon** Roemer & Schultes | 0 / 1 / 30 | |
| **Tripsacum** L. | 1 / 1 / 9 | intro. cult. |
| **Triraphis** R. Br. | 0 / 1 / 6 | |
| **Trisetum** Pers. | 0 / 1 / 75 | |
| Tristegis = Melinis | | |
| Urochloa see Alloteropsis, Brachiaria, Panicum | | |
| **Vetiveria** Bory | 0 / 1 / 2 | |
| **Zea** L. | 1 / 1 / 4 | intro. cult. ed. |
| **Zoysia** Willd. | 2 / 2 / 6 | |

## PODOCARPACEAE

0(20) / 12(155)

(see also Phyllocladaceae)

| | |
|---|---|
| **Dacrycarpus** (Endl.) Laubenf. (Podocarpus) | 0 / 1 / 9 |
| **Dacrydium** Lambert | 2 / 4 / 25 |
| **Decussocarpus** Laubenf. (Nageia) | 0 / 1 / 8 |
| **Falcatifolium** Laubenf. (Podocarpus) | 0 / 1 / 5 |
| Nageia see Decussocarpus | |
| **Podocarpus** L'Hérit. ex Pers. | 8 / 12 / 94 |
| Prumnopitys = Podocarpus | |
| Retrophyllum see Decussocarpus | |
| Stachycarpus = Podocarpus | |

## PODOSTEMACEAE

0(2) / 50(275)

| | |
|---|---|
| **Podostemum** Michaux | 0 / 1 / 18 |
| **Torrenticola** Domin ex Steenis | 0 / 1 / 1 |

## POLYGALACEAE

5(17) / 18(950)

(see also Xanthophyllaceae)

| | | |
|---|---|---|
| **Bredemeyera** Willd. (Polygala) | 0 / 1 / 20 | |
| Comesperma Labill. (Bredemeyera) | no record | |
| Epirixanthes = Salomonia | | |
| **Eriandra** P. Royen & Steenis | 1 / 1 / 1 | endemic in Papuasia |
| **Polygala** L. | 2 / 12 / 500 | |
| **Salomonia** Lour. | 2 / 2 / 8 | |

| | | |
|---|---|---|
| Securidaca L. | 0 / 1 / 80 | |

### POLYGONACEAE

| | | |
|---|---|---|
| | 5(20) / 51(1050) | |
| Antigonon Endl. | 0 / 1 / 3 | intro. |
| Calacinum see Muehlenbeckia | | |
| Coccoloba P. Browne (Muehlenbeckia) | 1 / 1 / 150 | |
| Muehlenbeckia Meissner | 0 / 3 / 15 | |
| Polygonum L. | 1 / 13 / 150 | |
| Rumex L. | 0 / 2 / 200 | |

### PONTEDERIACEAE

| | | |
|---|---|---|
| | 2(3) / 7(29) | |
| Eichhornia Kunth | 1 / 1 / 7 | |
| Monochoria C. Presl | 1 / 2 / 3 | |
| Pontederia see Monochoria | | |

### PORTULACACEAE

| | | |
|---|---|---|
| | 3(7) / 22(400) | |
| Montia L. | 0 / 1 / 15 | |
| Portulaca L. | 1 / 4 / 40 | |
| Talinum Adans. | 0 / 2 / 50 | |

### POTAMOGETONACEAE

| | |
|---|---|
| | 1(5) / 2(90) |

(see also Cymodoceaeceae, Ruppiaceae, Zannichelliaceae)

| | |
|---|---|
| Potamogeton L. | 1 / 5 / 90 |

### PRIMULACEAE

| | |
|---|---|
| | 3(5) / 22(800) |
| Androsace L. (Primula) | 0 / 1 / 100 |
| Lysimachia L. | 0 / 3 / 150 |
| Primula L. | 0 / 1 / 400 |

### PROTEACEAE

| | | |
|---|---|---|
| | 8(71) / 75(1350) | |
| Banksia L. f. | 0 / 1 / 71 | |
| Bleasdalea see Gevuina | | |
| Embothrium see Oreocallis | | |
| Euplassa see Gevuina | | |
| Finschia Warb. | 1 / 4 / 4 | |
| Gevuina Molina | 0 / 1 / 3 | |
| Grevillea R. Br. ex J. Knight | 1 / 7 / 250 | |
| Helicia Lour. | 0 / 52 / 87 | |
| Macadamia F. Muell. | 0 / 2 / 9 | cult. econ. |
| Oreocallis R. Br. | 0 / 2 / 5 | |
| Stenocarpus R. Br. | 0 / 2 / 22 | |

### PUNICACEAE

| | | |
|---|---|---|
| | 1(1) / 1(2) | |
| Punica L. | 0 / 1 / 2 | cult. ed. |

### RANUNCULACEAE

| | |
|---|---|
| | 3(43) / 58(1750) |
| Clematis L. | 2 / 7 / 230 |

**Ranunculus** L.     0 / 35 / 250

**Thalictrum** L.     0 / 1 / 85

## RESTIONACEAE     2(2) / 38(400)

**Leptocarpus** R. Br.     0 / 1 / 16

**Restio** Rottb.     0 / 1 / 88

Rafflesiaceae see Mitrastemmataceae

## RHAMNACEAE     11(33) / 53(875)

**Alphitonia** Reisseck ex Endl.     2 / 5 / 6

**Berchemia** Necker ex DC.     0 / 1 / 12

**Colubrina** Rich. ex Brongn.     1 / 2 / 31

**Dallachya** F. Muell. (Rhamnella)     0 / 1 / 1

**Emmenosperma** F. Muell.     0 / 1 / 3

**Gouania** Jacq.     1 / 4 / 26

**Maesopsis** Engl.     1 / 1 / 1     intro.

**Rhamnus** L.     0 / 7 / 125

**Smythea** Seemann ex A. Gray     2 / 4 / 7

**Ventilago** Gaertner     0 / 2 / 35

**Ziziphus** Miller (Zizyphus)     1 / 5 / 86

## RHIZOPHORACEAE     7(22) / 16(130)

**Bruguiera** Lam.     3 / 6 / 8

**Carallia** Roxb.     1 / 4 / 10

**Ceriops** Arn.     1 / 2 / 2

**Crossostylis** Forster & Forster f.     2 / 3 / 10

**Gynotroches** Blume     1 / 1 / 1

**Kandelia** (DC.) Wight & Arn.     0 / 1 / 1

**Rhizophora** L.     3 / 5 / 9

## ROSACEAE     7(57) / 107(3100)

(see also Chrysobalanaceae)

**Acaena** Mutis ex L.     0 / 2 / 100

Ancistrum = Acaena

**Fragaria** L.     0 / 1 / 12     intro. cult. ed.

**Potentilla** L.     0 / 12 / 500

**Prunus** L.     2 / 15 / 400

Pygeum = Prunus

**Rosa** L.     0 / 1 / 100     intro. cult. orn.

**Rubus** L.     4 / 25 / 250

**Spiraea** L.     0 / 1 / 70

Roxburghiaceae see Stemonaceae

## RUBIACEAE     73(838) / 637(10700)

Adenosacme = Mycetia

Adina see Metadina

**Airosperma** Schumann & Lauterb. — 0 / 4 / 6

**Amaracarpus** Blume — 2 / 49 / 60

**Anotis** DC. (Neanotis) — 0 / 1 / 31

**Anthocephalus** A. Rich. — 1 / 2 / 2 — intro. cult.

**Anthorrhiza** Huxley & Jebb — 0 / 8 / 8 — endemic in New Guinea

Antirhea see Guettardella

**Argostemma** Wallich — 0 / 23 / 100

**Badusa** A. Gray — 0 / 1 / 3

**Bikkia** Reinw. — 3 / 5 / 20

Borreria see Spermacoce

**Calycosia** A. Gray — 1 / 2 / 5

**Canthium** Lam. (Plectronia) — 3 / 15 / 50

Carinta = Geophila

**Cephaelis** Sw. — 1 / 5 / 100

**Chaetostachydium** Airy Shaw — 0 / 3 / 3 — endemic in New Guinea

Chaetostachys = Chaetostachydium

**Chassalia** Comm. ex Poiret — 0 / 2 / 42

Chomelia = Tarenna

Cinchona see Badusa

**Coelopyrena** Valeton — 0 / 1 / 1 — ident.?

**Coelospermum** Blume (Caelospermum) — 0 / 3 / 15

**Coffea** L. — 1 / 5 / 40 — cult. econ.

**Coprosma** Forster & Forster f. — 0 / 10 / 90

Coptophyllum Korth. — no record

**Coptosapelta** Korth. — 0 / 5 / 13

**Dentella** Forster & Forster f. — 0 / 2 / 10

**Diodia** L. — 0 / 1 / 30

**Diplospora** DC. (Tricalysia) — 0 / 1 / 25

Dolianthus = Amaracarpus

**Dolicholobium** A. Gray — 9 / 25 / 28

Fagraeopsis = Mastixiodendron

Gaertnera Lam. — no record

**Galium** L. — 0 / 10 / 400

**Gardenia** Ellis — 3 / 15 / 200

Geocardia = Geophila

**Geophila** D. Don — 3 / 3 / 20

Gomozia = Nertera

**Gouldia** A. Gray (Psychotria) — 0 / 1 / 3

Grumilea = Psychotria

**Guettarda** L. — 1 / 3 / 60

**Guettardella** Champ. ex Benth. (Anthirea) — 1 / 2 / 20

**Hedyotis** L. — 8 / 17 / 150

**Hydnophytum** Jack — 9 / 56 / 60

| | | |
|---|---|---|
| Hypobathrum Blume | no record | |
| **Ixora** L. | **5 / 44 / 400** | 1 intro. cult. orn. |
| **Kajewskiella** Merr. & Perry | **1 / 2 / 2** | endemic in Papuasia |
| **Knoxia** L. | **1 / 1 / 15** | |
| **Lasianthus** Jack | **2 / 17 / 150** | |
| Lasiostoma see Strychnos (Loganiaceae) | | |
| **Litosanthes** Blume | **0 / 3 / 5** | |
| **Lucinaea** DC. | **0 / 10 / 25** | |
| Mapouria Aublet | no record | |
| **Maschalodesme** Schumann & Lauterb. | **0 / 2 / 2** | endemic in New Guinea |
| **Mastixiodendron** Melchior | **2 / 4 / 7** | |
| **Metabolos** Blume (Hedyotis) | **0 / 1 / 10** | |
| **Metadina** Bakh. f. | **0 / 1 / 1** | |
| **Mitracarpus** Zucc. (Mitracarpum) | **0 / 1 / 30** | |
| **Mitragyna** Korth. | **0 / 2 / 10** | |
| **Morinda** L. | **5 / 16 / 50** | |
| **Mussaenda** L. | **5 / 31 / 200** | |
| **Mycetia** Reinw. | **1 / 1 / 25** | |
| **Myrmecodia** Jack | **1 / 26 / 45** | |
| Myrmedoma = Myrmephytum | | |
| **Myrmephytum** Becc. | **0 / 5 / 8** | |
| **Nauclea** Merr. | **4 / 4 / 10** | |
| **Neonauclea** Merr. | **4 / 4 / 6** | |
| **Nertera** Banks & Sol. ex Gaertner | **0 / 3 / 6** | |
| **Oldenlandia** L. (Hedyotis) | **1 / 21 / 300** | |
| **Ophiorrhiza** L. | **7 / 52 / 150** | |
| Ourouparia = Uncaria | | |
| **Pachystylus** Schumann | **0 / 2 / 2** | endemic in New Guinea |
| **Paederia** L. | **0 / 1 / 20** | |
| **Pavetta** L. | **1 / 5 / 350** | |
| **Pentas** Benth. | **0 / 1 / 34** | cult. orn. |
| **Pertusadina** Ridsd. (Adina) | **0 / 1 / 4** | |
| Petunga = Hypobathrum | | |
| Plectronia see Canthium | | |
| Pogonolobus = Coelospermum | | |
| Polyphragmon = Timonius | | |
| **Psychotria** L. | **14 / 124 / 1400** | |
| **Randia** L. | **8 / 36 / 300** | |
| **Rhadinopus** S. Moore | **0 / 2 / 2** | endemic in New Guinea |
| **Richardia** L. | **0 / 1 / 15** | |
| **Saprosma** Blume | **2 / 8 / 10** | |
| Sarcocephalus see Nauclea | | |
| **Scyphiphora** Gaertner f. | **1 / 1 / 1** | |
| **Siphonandrium** Schumann | **0 / 1 / 1** | endemic in New Guinea |

| | | |
|---|---|---|
| **Spermacoce** L. | 5 / 5 / 100 | some nat. |
| Sphaerophora = Morinda | | |
| Stephegyne = Mitragyna | | |
| Stylocoryna = Tarenna | | |
| **Tarenna** Gaertner | 2 / 9 / 370 | |
| **Timonius** DC. | 11 / 71 / 150 | |
| **Tricalysia** A. Rich. ex DC. | 0 / 1 / 100 | |
| **Uncaria** Schreber | 10 / 10 / 34 | |
| Uragoga = Cephaelis | | |
| **Urophyllum** Jack ex Wallich | 1 / 16 / 150 | |
| Uruparia = Uncaria | | |
| Vangueria see Canthium | | |
| **Versteegia** Valeton | 2 / 6 / 6 | endemic in Papuasia |
| **Wendlandia** Bartling ex DC. | 0 / 4 / 56 | |
| **Xanthophytum** Reinw. ex Blume | 0 / 2 / 15 | |

## RUPPIACEAE

| | |
|---|---|
| | 1 / 1(7) |
| Ruppia L. | no record |

## RUTACEAE

| | | |
|---|---|---|
| | 27(167) / 161(1650) | |
| **Acronychia** Forster & G. Forster | 0 / 34 / 44 | |
| Androcephalium = Lunasia | | |
| **Atalantia** Corr. Serr. | 0 / 2 / 18 | |
| **Bauerella** Borzi | 0 / 1 / 2 | |
| **Bouchardatia** Baillon (Melicope) | 0 / 1 / 2 | |
| Chalcas = Murraya | | |
| **Citrus** L. | 7 / 7 / 16 | intro. cult. ed. |
| **Clausena** Burm. f. | 0 / 1 / 23 | |
| **Clymenia** Swingle | 0 / 2 / 2 | endemic in Papuasia |
| Coombea = Medicosma | | |
| Echinocitrus = Triphasia | | |
| **Euodia** J.R. & G. Forster (Melicope) | 11 / 37 / 45 | |
| Evodia = Euodia | | |
| **Evodiella** Van der Linden | 0 / 3 / 3 | |
| Fagara = Zanthoxylum | | |
| **Flindersia** R. Br. | 2 / 5 / 17 | intro. cult. |
| **Geijera** Schott | 0 / 1 / 7 | |
| **Glycosmis** Corr. Serr. | 0 / 5 / 50 | |
| **Halfordia** F. Muell. | 0 / 3 / 4 | |
| Herzogia see Melicope | | |
| Hormopetalum = Sericolea (Elaeocarpaceae) | | |
| Hunsteinia see Rapanea (Myrsinaceae) | | |
| Lamiofrutex = Vavaea (Meliaceae) | | |
| **Lunasia** Blanco | 0 / 1 / 10 | |
| **Luvunga** Buch.-Ham. ex Wight & Arn. | 0 / 2 / 12 | |

Medicosma Hook. f.     no record

Melanococca = Rhus (Anacardiaceae)

**Melicope** Forster & Forster f.     3 / 18 / 20

**Merope** M. Roemer (Atalantia)     0 / 1 / 1

**Microcitrus** Swingle     0 / 2 / 6

**Micromelum** Blume     2 / 3 / 10

**Monanthocitrus** Tanaka     0 / 3 / 3     endemic in New Guinea

**Murraya** Koenig ex L.     2 / 3 / 4

**Paramigyna** Wight (Atalantia)     0 / 1 / 20

**Poncirus** Raf.     1 / 1 / 1     intro. cult. orn.

Terminthodia = Tetractomia

**Tetractomia** Hook. f. (Melicope)     1 / 6 / 6

**Triphasia** Lour.     1 / 2 / 3

**Wenzelia** Merr.     1 / 7 / 9

**Zanthoxylum** L. (Xanthoxylum)     3 / 13 / 200

## SABIACEAE     2(5) / 3(48)

**Meliosma** Blume     0 / 4 / 25

Millingtonia see Meliosma

**Sabia** Colebr.     1 / 1 / 19

## SALVADORACEAE     1(1) / 2(11)

**Azima** Lam.     0 / 1 / 4

## SANSEVIERACEAE     1(2) / 1(12)

**Sanseviera** Thunb.     1 / 2 / 12     nat. cult. orn.

## SANTALACEAE     9(43) / 36(500)

**Anthobolus** R. Br. (Exocarpos)     0 / 1 / 3

**Cladomyza** Danser     0 / 16 / 20

**Dendromyza** Danser     2 / 7 / 7

**Dendrotrophe** Miq.     0 / 3 / 4

**Exocarpos** Labill.     0 / 5 / 26

Henslowia = Dendrotrophe

**Hylomyza** Danser (Dufrenoia)     0 / 4 / 6

**Phacellaria** Benth.     0 / 1 / 7

**Santalum** L.     0 / 4 / 9

Scleromelum = Scleropyrum

**Scleropyrum** Arn.     0 / 2 / 6

## SAPINDACEAE     33(171) / 145(1325)

**Alectryon** Gaertner     0 / 11 / 15

**Allophylus** L.     3 / 2 / 175

Aphania = Lepisanthes

Aporetica = Allophylus

**Arytera** Blume     2 / 6 / 25

| | | |
|---|---|---|
| **Atalaya** Blume | 0 / 1 / 11 | |
| **Cardiospermum** L. | 1 / 2 / 14 | |
| Crossonephelis = Glenniea | | |
| Cubilia Blume | no record | |
| Cupania see Arytera, Nephelium | | |
| **Cupaniopsis** Radlk. | 2 / 20 / 66 | |
| **Dictyoneura** Blume | 0 / 2 / 2 | |
| **Dimocarpus** Lour. (Litchi) | 0 / 1 / 6 | |
| **Diploglottis** Hook. f. | 0 / 1 / 3 | |
| **Dodonaea** Miller | 1 / 2 / 50 | |
| **Elattostachys** Radlk. | 1 / 4 / 14 | |
| Erioglossum = Lepisanthes | | |
| Euphoria = Dimocarpus | | |
| Euphorianthus = Diploglottis | | |
| **Ganophyllum** Blume | 1 / 1 / 1 | |
| **Glenniea** Hook. f. | 0 / 1 / 8 | |
| **Guioa** Cav. | 2 / 28 / 40 | |
| **Harpullia** Roxb. | 10 / 30 / 26 | |
| Irina = Pometia | | |
| **Jagera** Blume | 0 / 3 / 4 | |
| **Lepiderema** Radlk. | 0 / 3 / 8 | |
| **Lepidopetalum** Blume | 2 / 5 / 6 | |
| **Lepisanthes** Blume | 0 / 6 / 24 | |
| **Litchi** Sonn. | 0 / 1 / 1 | cult. ed. |
| **Mischocarpus** Blume | 1 / 9 / 11 | |
| Mischocodon = Mischocarpus | | |
| **Nephelium** L. | 0 / 1 / 35 | |
| **Pometia** Forster & Forster f. | 2 / 2 / 2 | |
| **Rhysotoechia** Radlk. | 0 / 4 / 14 | |
| **Sapindus** L. | 0 / 1 / 13 | |
| **Sarcopteryx** Radlk. | 0 / 7 / 8 | |
| **Sarcotoechia** | 0 / 4 / 4 | |
| **Schleichera** Willd. | 0 / 1 / 1 | |
| Schmidelia = Allophylus | | |
| Spanoghea = Alectryon | | |
| **Synima** Radlk. | 0 / 1 / 1 | |
| **Toechima** Radlk. | 1 / 3 / 8 | |
| **Trigonachras** Radlk. | 0 / 1 / 9 | |
| **Tristira** Radlk. | 0 / 1 / 4 | ident.? |
| **Tristiropsis** Radlk. | 4 / 6 / 8 | |

## SAPOTACEAE    13(100) / 116(1100)

Achras = Manilkara
Albertisiella = Pouteria

Bassia = Madhuca
Beauvisagea = Pouteria
Beccariella = Pichonia
**Burckella** Pierre                          7 / 9 / 11
Bureavella = Pouteria
Cassidispermum = Burckella
Chelonespermum = Burckella
**Chrysophyllum** L.                          2 / 4 / 80
Ganua = Madhuca
Illipe = Madhuca
Krausella = Pouteria
Lucuma = Pouteria
**Madhuca** J. Gmelin                         1 / 5 / 85
**Magodendron** Vink                          0 / 1 / 1        endemic in New Guinea
**Manilkara** Adans.                          2 / 6 / 70            1 intro. cult.
**Mimusops** L.                               1 / 2 / 57
Niemeyera see Chrysophyllum and Dysoxylum (Meliaceae)
**Nortuia** Hook. f.                          0 / 1 / 1
**Palaquium** Blanco                          8 / 19 / 115
**Payena** A. DC.                             0 / 2 / 16
**Pichonia** Pierre                           0 / 6 / 100
Planchonella see Pichonia, Pouteria
**Pouteria** Aublet                           3 / 20 / 80
**Sarcosperma** Hook. f.                      0 / 1 / 6
**Sideroxylon** L. (Planchonella)            1 / 24 / 100
Woikoia (Wokoia) = Pouteria

Sarcospermataceae see Sapotaceae

Saurauiaceae see Actinidiaceae

## SAXIFRAGACEAE                              2(12) / 37(475)

(see also Cunoniaceae, Grossulariaceae, Hydrangeaceae, Myrtaceae)
**Astilbe** Buch.-Ham. ex D. Don             0 / 2 / 12
Dedea = Quintinia
**Quintinia** A. DC.                          0 / 10 / 14

## SCROPHULARIACEAE                          23(88) / 222(4500)
**Adenosma** R. Br.                           0 / 5 / 15
**Angelonia** Bonpl.                          0 / 2 / 25
**Artanema** D. Don                           0 / 1 / 4
**Asarina** Miller                            no record
**Bacopa** Aublet                             0 / 1 / 56
Bonnaya see Lindernia
**Buchnera** L.                               0 / 5 / 100

Capraria see Lindernia
**Centranthera** R. Br.                                   0 / 2 / 9
**Detzneria** Schltr. ex Diels                           0 / 2 / 2          endemic in New Guinea
**Ellisiophyllum** Maxim.                                0 / 1 / 1
Erinus see Poarium
**Euphrasia** L.                                         0 / 10 / 450
Gratiola see Lindernia
**Hebe** Comm. ex Juss. (Veronica, Parahebe)             0 / 2 / 75
Herpestis = Bacopa
**Ilysanthes** Raf. (Lindernia)                          0 / 1 / 50
**Limnophila** R. Br.                                    1 / 6 / 36
Lindenbergia Lehm.                                       no record
**Lindernia** All.                                       1 / 14 / 50
Maurandya = Asarina
**Mazus** Lour.                                          0 / 2 / 30
Mella = Bacopa
Mimulus see Torenia
**Parahebe** W. Oliver                                   0 / 12 / 30
**Poarium** Desv. (Stemodia)                             0 / 1 / 20
**Rhamphicarpa** Benth.                                  0 / 1 / 6
**Russelia** Jacq.                                       1 / 1 / 52          intro. cult. orn.
**Scoparia** L.                                          0 / 1 / 20
**Sopubia** Buch.-Ham. ex D. Don                         0 / 1 / 60
**Striga** Lour.                                         0 / 8 / 40
**Torenia** L.                                           0 / 6 / 40
Vandellia = Lindernia
**Veronica** L.                                          0 / 3 / 250

## SIMAROUBACEAE                                         6(6) / 22(170)

(see also Surianaceae)
**Ailanthus** Desf.                                      1 / 1 / 5
**Brucea** J.F. Miller                                   0 / 1 / 6
Cardiocarpus = Soulamea
Cardiophora = Soulamea
**Harrisonia** R. Br. ex A. Juss.                        1 / 1 / 4
**Picrasma** Blume                                       1 / 1 / 8
**Quassia** L.                                           1 / 1 / 35
Samadera = Quassia
**Soulamea** Lam.                                        1 / 1 / 9

Siphonodontaceae see Celastraceae

## SMILACACEAE                                           4(23) / 10(225)
**Eustrephus** R. Br.                                    1 / 1 / 1
**Geitonoplesium** Cunn. ex R. Br.                       1 / 1 / 1

**Luzuriaga** Ruíz & Pavón     0 / 2 / 3
**Smilax** L.     3 / 19 / 200

## SOLANACEAE     10(88) / 90(2600)
**Brachistus** Miers     1 / 1 / 3
**Browallia** L.     0 / 1 / 6
Brugmansia Pers. (Datura)     no record
**Capsicum** L.     2 / 2 / 10     intro. cult. ed.
**Cestrum** L.     0 / 1 / 175     intro.
Cyphomandra C. Martius ex Sendtner     0 / 1 / 30     cult. ed.
**Datura** L.     1 / 2 / 8     nat.
Lycianthes see Solanum
**Lycopersicon** Miller (Solanum)     1 / 1 / 7     intro. cult. ed.
**Nicotiana** L.     1 / 1 / 67     intro. cult. econ.
Parascopolia see Solanum
**Physalis** L.     1 / 3 / 80
**Solanum** L.     14 / 75 / 1400     2 intro. cult.

## SONNERATIACEAE     2(4) / 2(7)
**Duabanga** Buch.-Ham.     0 / 1 / 2
**Sonneratia** L. f.     3 / 3 / 5

## SPARGANIACEAE     1(1) / 1(12)
**Sparganium** L.     0 / 1 / 12

## SPHENOCLEACEAE     1(1) / 1(2)
**Sphenoclea** Gaertner     1 / 1 / 2

Sphenostemonaceae see Aquifoliaceae

## STACKHOUSIACEAE     1(1) / 3(28)
**Stackhousia** Sm.     0 / 1 / 25

## STAPHYLEACEAE     1(1) / 5(27)
Kaernbachia = Turpinia
Staphylea L.     no record
**Turpinia** Vent.     0 / 1 / 10

## STEMONACEAE     1(2) / 4(32)
Roxburghia = Stemona
**Stemona** Lour.     0 / 5 / 25

## STERCULIACEAE     19(63) / 73(1500)
Abroma = Ambroma
**Ambroma** L. f.     1 / 2 / 2
Argyrodendron F. Muell. (Heritiera)     no record
**Brachychiton** Schott & Endl.     0 / 2 / 30
**Commersonia** Forster & Forster f.     1 / 1 / 10

| | | |
|---|---|---|
| **Firmiana** Marsili | 0 / 1 / 9 | |
| **Helicteres** L. | 0 / 1 / 40 | |
| **Heritiera** Dryander | 4 / 9 / 30 | |
| **Keraudrenia** Gay | 0 / 1 / 9 | |
| **Kleinhovia** L. | 1 / 1 / 1 | |
| **Leptonychia** Turcz. | 0 / 1 / 30 | |
| **Melhania** Forssk. | 0 / 1 / 60 | |
| **Melochia** L. | 3 / 5 / 54 | |
| **Pentapetes** L. | 0 / 1 / 1 | |
| **Pterocymbium** R. Br. | 0 / 5 / 15 | |
| **Pterospermum** Schreber | 0 / 1 / 25 | |
| **Pterygota** Schott & Endl. | 0 / 2 / 15 | |
| Scaphium Schott & Endl. | no record | |
| Seringia see Keraudrenia | | |
| **Sterculia** L. | 6 / 26 / 200 | |
| Tarrietia = Heritiera | | |
| **Theobroma** L. | 1 / 1 / 20 | intro. cult. econ. |
| **Triplochiton** Schumann | 1 / 1 / 3 | intro. cult. orn. |
| **Waltheria** L. | 0 / 1 / 67 | |
| **STILAGINACEAE** | 1(32) / 1(160) | |
| **Antidesma** L. | 7 / 32 / 160 | |
| **STRELITZIACEAE** | 2(2) / 3(7) | |
| **Ravenala** Adans. | 1 / 1 / 1 | intro. cult. orn. |
| **Strelitzia** Banks ex Dryander | 1 / 1 / 5 | intro. cult. orn. |
| **STYLIDIACEAE** | 1(1) / 5(170) | |
| Candollea = Stylidium | | |
| **Stylidium** Sw. ex Willd. | 0 / 1 / 150 | |
| **STYRACACEAE** | 2(2) / 12(165) | |
| **Bruinsmia** Boerl. & Koord. | 0 / 1 / 2 | |
| **Styrax** L. | 1 / 1 / 120 | |
| **SURIANACEAE** | 1(1) / 4(5) | |
| **Suriana** L. | 0 / 1 / 1 | |
| Symphoremataceae see Verbenaceae | | |
| **SYMPLOCACEAE** | 1(40) / 1(250) | |
| **Symplocos** Jacq. | 2 / 40 / 250 | |
| **TACCACEAE** | 1(3) / 1(10) | |
| **Tacca** Forster & Forster f. | 3 / 3 / 10 | |
| Ternstroemiaceae see Theaceae | | |
| Tetragoniaceae see Aizoaceae | | |

Tetramelaceae see Datiscaceae

## THEACEAE                          6(45) / 28(520)

**Adinandra** Jack                    0 / 2 / 70
**Archboldiodendron** Kobuski         0 / 2 / 2          endemic in New Guinea
Cyclandra = Ternstroemia
**Eurya** Thunb.                      3 / 26 / 70
**Gordonia** Ellis                    0 / 1 / 70
Laplacea = Gordonia
**Ploiarium** Korth.                  0 / 1 / 3
**Ternstroemia** Mutis ex L. f.       0 / 13 / 85
Tremanthera = Saurauia (Actinidiaceae)

Thunbergiaceae see Acanthaceae

## THYMELAEACEAE                      9(34) / 50(720)

**Aquilaria** Lam.                    0 / 1 / 15
Brachythalamus = Gyrinops
Dais see Phaleria
**Drapetes** Lam.                     1 / 1 / 4
Drimyspermum = Phaleria
**Enkleia** Griffith                  0 / 1 / 3
**Gonystylus** Teijsm. & Binnend.     2 / 3 / 20
**Gyrinops** Gaertner                 0 / 5 / 8
**Kelleria** Endl. (Drapetes)         0 / 3 / 7
**Phaleria** Jack                     3 / 14 / 20
**Pimelea** Banks & Sol. ex Gaertner  0 / 2 / 80
**Wikstroemia** Endl.                 2 / 4 / 70

## TILIACEAE                         9(54) / 48(725)

Althoffia = Trichospermum
**Berrya** Roxb.                      0 / 1 / 5
**Brownlowia** Roxb.                  1 / 2 / 25
**Colona** Cav.                       2 / 4 / 30
Columbia = Colona
**Corchorus** L.                      0 / 3 / 40          cult. econ.
**Eleutherostylis** Burret            0 / 1 / 1       endemic in New Guinea
Glabraria = Brownlowia
**Grewia** L.                         0 / 11 / 150
**Microcos** L.                       1 / 4 / 53
**Trichospermum** Blume               11 / 19 / 39
**Triumfetta** L.                     4 / 8 / 100

## TRIMENIACEAE                      2(4) / 2(5)

**Piptocalyx** Oliver ex Benth. (Trimenia)   0 / 1 / 2
**Trimenia** Seemann                  0 / 3 / 3

Triplostegiaceae see Valerianaceae

## TRIURIDACEAE                      1(33) / 6(42)
Andruris = Sciaphila
**Sciaphila** Blume                 3 / 33 / 33

## TROPAEOLACEAE                     0 / 3(88)
Tropaeolum L.                        no record

## TURNERACEAE                       1(1) / 10(110)
**Turnera** L.                       0 / 1 / 60          intro. cult. orn.

## TYPHACEAE                         1(2) / 1(12)
**Typha** L.                         0 / 2 / 12

## ULMACEAE                          5(18) / 16(140)
**Aphananthe** Planchon              1 / 1 / 5
**Celtis** L.                        10 / 6 / 60
**Gironniera** Gaudich.              3 / 4 / 6
**Parasponia** Miq.                  2 / 4 / 5
Solenostigma = Celtis
Sponia = Trema
**Trema** Lour.                      3 / 3 / 15

Umbelliferae see Apiaceae

## URTICACEAE                        22(222) / 52(1050)

(see also Cecropiaceae)
**Boehmeria** Jacq.                  4 / 1 / 50
Conocephalus = Poikilospermum (Cecropiaceae)
**Cypholophus** Wedd.                2 / 28 / 30
**Debregeasia** Gaudich.             0 / 2 / 5
**Dendrocnide** Miq.                 8 / 12 / 36
Distemon see Neodistemon
**Elatostema** Forster & Forster f.  16 / 57 / 200
Fleurya = Laportea
**Gibbsia** Rendle                   0 / 2 / 2          endemic in New Guinea
**Girardinia** Gaudich.              0 / 1 / 2          coll. uncertain
Gonostegia = Pouzolzia
**Hyrtanandra** Miq. (Pouzolzia)     0 / 1 / 15         coll. uncertain
**Laportea** Gaudich.                3 / 1 / 22
**Leucosyke** Zoll. & Moritzi        4 / 10 / 35
**Maoutia** Wedd.                    3 / 6 / 15
Memorialis = Hyrtanandra
Missiessya = Leucosyke
**Neodistemon** Babu & A.N. Henry    0 / 1 / 1
**Nothocnide** Blume                 1 / 4 / 5

**Oreocnide** Miq. (Villebrunea)          0 / 1 / 20
**Parietaria** L.                         0 / 1 / 20
**Pellionia** Gaudich. (Elatostema)       1 / 37 / 50
**Pilea** Lindley                         1 / 31 / 250
**Pipturus** Wedd.                        2 / 12 / 40
**Pouzolzia** Gaudich.                    3 / 4 / 50
**Procris** Comm. ex Juss.                3 / 6 / 20
Pseudopipturus = Nothocnide
Robinsoniodendron = Maoutia
Schychowskya = Laportea
Sciophila = Procris
**Urtica** L.                             0 / 1 / 45
**Villebrunea** Gaudich. ex Wedd.         0 / 3 / 8

## VALERIANACEAE                          1(2) / 17(400)
**Triplostegia** Wallich ex DC.           0 / 2 / 2

## VERBENACEAE                            16(124) / 90(1900)
**Archboldia** E. Beer & H.J. Lam         0 / 1 / 1        endemic in New Guinea
**Callicarpa** L.                         2 / 13 / 140
**Clerodendrum** L.                       6 / 36 / 400
Cornutia see Premna
**Duranta** L.                            0 / 1 / 30       intro. cult. orn.
**Faradaya** F. Muell.                    2 / 13 / 17
Geunsia = Callicarpa
**Glossocarya** Wallich ex Griffith       0 / 1 / 9
Gmelina L.                                3 / 11 / 35      1 intro. cult.
**Lantana** L.                            1 / 2 / 150
Lippia see Phyla
**Petraeovitex** Oliver                   1 / 2 / 7
**Phyla** Lour. (Lippia)                  1 / 1 / 15
**Premna** L.                             4 / 18 / 200
Pygmaeopremna = Premna
Sphenodesme Jack                          no record
**Stachytarpheta** Vahl                   4 / 4 / 65
**Tectona** L. f.                         1 / 1 / 4        intro. cult. econ.
**Teijsmanniodendron** Koord.             2 / 3 / 14
**Verbena** L.                            0 / 2 / 250
**Vitex** L.                              4 / 14 / 250
Viticipremna = Vitex
Xerocarpa = Teijsmanniodendron

## VIOLACEAE                              4(15) / 23(930)
**Agatea** A. Gray                        1 / 1 / 1
Alsodeia = Rinorea

| | | |
|---|---|---|
| **Hybanthus** Jacq. | 0 / 2 / 150 | |
| Ionidium = Hybanthus | | |
| Pentaloba = Rinorea | | |
| **Rinorea** Aublet | 4 / 4 / 200 | |
| **Viola** L. | 1 / 8 / 500 | intro. cult. orn. |
| **VISCACEAE** | 4(16) / 8(450) | |
| **Ginalloa** Korth. | 0 / 1 / 15 | |
| **Korthalsella** Tieghem | 0 / 1 / 45 | |
| **Notothixos** Oliver | 1 / 7 / 8 | |
| **Viscum** L. | 0 / 7 / 100 | |
| **VITACEAE** | 5(52) / 13(800) | |
| **Ampelocissus** Planchon | 0 / 3 / 95 | |
| **Cayratia** Juss. | 3 / 7 / 45 | |
| **Cissus** L. | 1 / 24 / 350 | |
| **Tetrastigma** (Miq.) Planchon | 3 / 17 / 90 | |
| **Vitis** L. | 0 / 1 / 65 | |
| **WINTERACEAE** | 2(20) / 5(60) | |
| Belliolum = Zygogynum | | |
| Bubbia = Zygogynum | | |
| **Drimys** Forster & Forster f. | 1 / 1 / 9 | |
| Tasmannia = Drimys | | |
| Tetrathalamus see Drimys | | |
| **Zygogynum** Baillon | 4 / 19 / 30 | |
| **XANTHOPHYLLACEAE** | 1(5) / 1(93) | |
| **Xanthophyllum** Roxb. | 1 / 5 / 93 | |
| **XANTHORRHOEACEAE** | 1(2) / 9(60) | |
| **Lomandra** Labill. | 0 / 2 / 35 | |
| Xerotes = Lomandra | | |
| **XYRIDACEAE** | 1(6) / 5(260) | |
| **Xyris** L. | 0 / 6 / 240 | |
| **ZAMIACEAE** | 0 / 8(85) | |
| **Macrozamia** Miq. | no record | |
| Zannichellaceae see Cymodoceaceae | | |
| **ZINGIBERACEAE** | 16(207) / 53(1200) | |
| Achasma = Amomum | | |
| **Alpinia** Roxb. | 13 / 86 / 250 | 1 intro. cult. orn. |
| **Amomum** Roxb. | 1 / 8 / 90 | |
| **Boesenbergia** Kuntze | 0 / 1 / 20 | ident.? |
| Catimbium = Alpinia | | |
| **Costus** L. | 2 / 3 / 90 | |

| | | |
|---|---|---|
| **Curcuma** Roxb. | 1 / 5 / 40 | intro. cult. econ. |
| **Elettaria** Maton | 1 / 1 / 6 | intro. cult. econ. |
| **Elettariopsis** Baker | 0 / 1 / 10 | |
| Eriolopha = Alpinia | | |
| Etlingera = Amomum | | |
| Geanthus = Amomum | | |
| **Globba** L. | 1 / 10 / 70 | |
| Guillainea = Alpinia | | |
| Haplochorema see Kaempferia | | |
| **Hedychium** J. Koenig | 0 / 2 / 50 | |
| Hellenia see Alpinia, Costus | | |
| Hellwigia = Alpinia | | |
| **Hornstedtia** Retz. | 1 / 3 / 60 | |
| **Kaempferia** L. | 0 / 1 / 50 | |
| Nanochilus Schumann | no record | |
| Naumannia = Riedelia | | |
| **Nicolaia** Horan. (Etlingera) | 0 / 2 / 8 | |
| Phaeomeria = Nicolaia | | |
| Psychanthus = Alpinia | | |
| **Riedelia** Oliver | 1 / 62 / 80 | |
| **Tapeinochilos** Miq. | 1 / 13 / 15 | |
| **Thylacophora** Ridley (Riedelia) | 0 / 1 / 1 | endemic in New Guinea |
| **Zingiber** Boehmer | 1 / 6 / 85 | nat. cult. econ. |

## ZYGOPHYLLACEAE

1(1) / 27(250)

| | | |
|---|---|---|
| **Tribulus** L. | 0 / 1 / 25 | |

# III. Some vernacular names and their scientific equivalents

| | |
|---|---|
| aiai | Eugenia malaccensis |
| aibika | Abelmoschus manihot |
| aila | Inocarpus fagiferus |
| aitan | Alstonia scholaris |
| akas | Acacia spp. |
| alang alang | Imperata cylindrica |
| amberoi | Pterocymbium beccarii |
| arang | Pandanus spp. |
| arrowroot | Maranta spp. |
| arurut | Maranta spp. |
| aupa | Amaranthus tricolor |
| avocado | Persea americana |
| baibai | Cycas circinalis |
| balsa | Ochroma lagopus |
| bamboo | Arundinaria spp. |
| | Bambusa spp. |
| | Dinochloa sp. |
| | Gigantochloa sp. |
| | Oxytenanthera spp. |
| | Schizostachyum spp. |
| banana | Musa spp. |
| banyan tree | Ficus spp. |
| basswood | Endospermum medullosum |
| bata | Persea americana |
| beech | Nothofagus spp. |
| betel nut | Areca catechu |
| betel pepper | Piper betle |
| bin | Phaseolus spp. |
| bitum | Vitex spp. |
| bottle gourd | Lagenaria siceraria |
| boxwood | Xanthophyllum papuanum |
| bread fruit | Artocarpus spp. |
| brus | Nicotiana tabacum |
| buai | Areca catechu |
| bukbuk | Dysoxylum caulostachyum |
| Bulolo ash | Hibiscus papuodendron |
| busu plum | Maranthes corymbosa |
| cabbage | Brassica spp. |
| camphorwood | Cinnamomum spp. |
| candle nut | Aleurites moluccana |
| carambola | Averrhoa carambola |
| cardamom | Elettaria cardamomum |
| carrot | Daucus carota |

| | |
|---|---|
| cashew | Anacardium occidentale |
| cassava | Manihot esculenta |
| castor oil | Ricinus communis |
| celery-top pine | Phyllocladus hypophyllus |
| chilli | Capsicum spp. |
| coachwood | Ceratopetalum succirubrum |
| cocoa | Theobroma cacao |
| coffee | Coffea spp. |
| coral tree | Erythrina indica |
| coriander | Coriandrum sativum |
| corn | Zea mays |
| cowpea | Vigna unguiculata |
| crab apple | Schizomeria serrata |
| cucumber | Cucumis spp. |
| custard apple | Annona squamosa |
| cypress | Papuacedrus papuanus |
| daka | Piper betle |
| drai | Cocos nucifera (dry) |
| drip | Cocos nucifera (young) |
| durian | Durio zibethinus |
| ebony | Diospyros spp. |
| egg plant | Solanum melongena |
| erima | Octomeles sumatrana |
| faivkorn | Averrhoa carambola |
| fig | Ficus spp. |
| galip | Canarium indicum |
| garamut | Vitex cofassus |
| garawa | Anisoptera polyandra |
| garlic | Allium sativum |
| garo-garo | Mastixiodendron pachyclados |
| ginger | Zingiberaceae gen. |
| gorgor | Alpinia spp. |
| | Tapeinochilus spp. |
| cotton | Gossypium spp. |
| gourd | Trichosanthes sp. |
| granadillo | Passiflora spp. |
| green pepper | Capsicum spp. |
| grey milkwood | Cerbera floribunda |
| groundnut | Arachis hypogaea |
| guava | Psidium guajava |
| hard yar | Casuarina papuana |
| hebsen | Pisum sativum |
| hoop pine | Araucaria cunninghamii |

| | | | |
|---|---|---|---|
| horse-radish tree | Moringa oleifera | lettuce | Lactuca sativa |
| hot pepper | Capsicum spp. | limbum | Areca sp. |
| ironwood | Intsia bijuga | | Caryota sp. |
| ivorywood | Siphonodon celastrineus | | Kentiopsis archontophoenix |
| Japanese cherry | Muntingia calabura | litchi | Litchi chinensis |
| Java cedar | Bischofia javanica | lombo | Capsicum spp. |
| kakao | Theobroma cacao | mais | Zea mays |
| kalopilum | Calophyllum inophyllum | maize | Zea mays |
| kambang | Trichosanthes sp. | mahagony | Swietenia spp. |
| kamerere | Eucalyptus deglupta | | Weinmannia spp. |
| kanda | Calamus spp. | makas | Hibiscus spp. |
| kango | Nasturtium officinale | malas | Homalium foetidum |
| kanu | Alstonia scholaris | Malay apple | Eugenia malaccensis |
| kapiak | Artocarpus spp. | mambu | Arundinaria spp. |
| kapok | Bombax ceiba | | Bambusa spp. |
| | Ceiba pentandra | | Dinochloa sp. |
| karapua | Musa | | Gigantochloa sp. |
| karuka | Pandanus spp. | | Oxytenanthera spp. |
| kasang | Arachis hypogaea | | Schizostachyum spp. |
| kaswel | Ricinus communis | mami | Dioscorrea esculenta |
| kaukau | Ipomoea batatas | mandarin | Citrus nobilis |
| kaur | hard bamboo | mang bin | Vigna mungo |
| kauri pine | Agathis spp. | mangas | Hibiscus spp. |
| kawawar | Zingiber officinale | mango | Mangifera indica |
| kawiwi | Areca spp. | mangro | Rhizophora mucronata |
| | Howea belmoreana | manioc | Manihot esculenta |
| kempas | Koompassia grandiflora | mareo | Pometia pinnata |
| kerasin | Cordia subcordata | marihuana | Canabis sativa |
| kirsen | Aberia sp. Dougalis sp. | marita | Pandanus spp. |
| klinki pine | Araucaria hunsteinii | marmar | Samanea saman |
| kokomo | Chisocheton sp. | | Poinciana delnis |
| kokonas | Cocos nucifera | massoia | Cryptocarya spp. |
| kon | Zea mays | melen | Citrullus lanatus |
| koniak | Piper methysticum | mon | Dracontomelum mangiferum |
| kopi | Coffea spp. | muli | Citrus aurantifolia |
| kopra | Cocos nucifera (dried) | mung bean | Vigna mungo |
| kulau | Cocos nucifera (drinking) | nettle | Dendrocnide sp. |
| kunai | Imperata cylindrica | New Guinea creeper | Mucuna bennettii |
| kwila | Intsia bijuga | New Guinea walnut | Dracontomelum |
| labula | Anthocephalus cadamba | | mangiferum |
| laulau | Syzygium spp. | nar | Pterocarpus indicus |
| laup | Dracontomelum mangiferum | numa | Antiaris toxicaria |
| lemon grass | Cymbopogon citratus | nutmeg | Horsfieldia spp. |

## III. Some vernacular names and their scientific equivalents

| | | | |
|---|---|---|---|
| | Myristica spp. | shallot | Allium cepa |
| oak | Castanopsis acuminatissima | sida | Toona sureni |
| | Lithocarpus spp. | silkwood | Flindersia pimenteliana |
| oil palm | Elaeis guineensis | sirsen | Muntingia calabura |
| onion | Allium cepa | sisal | Agave sisalana |
| palpal | Erythrina indica | skin diwai | Cinnamomum spp. |
| papaya | Carica papaya | snake bean | Vigna unguiculata |
| paprika | Capsicum spp. | snek bin | Vigna unguiculata |
| Papua nutmeg | Myristica argentea | soursap | Annona muricata |
| passion fruit | Passiflora edulis | soya bean | Glycine max |
| pau | Barringtonia calyptocalyx | spakbrus | Canabis sativa |
| pawpaw | Carica papaya | spinach | Amaranthus spp. |
| peanut | Arachis hypogaea | stik masis | Spathodea campanulata |
| peas | Pisum sativum | stinging tree | Dendrocnide spp. |
| pencil cedar | Palaquium warburgianum | sugar cane | Saccharum officinarum |
| pigeon pea | Cajanus cajan | sugar palm | Arenga pinnata |
| pineapple | Ananas sativus | sweet basil | Ocimum basilicum |
| pitpit | Saccharum robustum | sweet potato | Ipomoea batatas |
| | S. spontaneum | switmuli | Citrus aurantium |
| pomelo | Citrus decumana | sword grass | Imperata cylindrica |
| popo | Carica papaya | Tahitian chestnut | Canarium indicum |
| potato | Solanum tuberosum | talis | Terminalia catappa |
| pumpkin | Cucurbita spp. | | T. complanata |
| purpur | Codiaeum variegatum | tanget | Cordyline spp. |
| quandong | Elaeocarpus sphaericus | | Taetsia fructicosa |
| raintree | Samanea saman | tapiok | Manihot esculenta |
| rais | Oryza sativa | taro | Colocasia esculenta |
| rattan | Calamus spp. | | Xanthosoma sagittifolia |
| red cedar | Toona surenii | taun | Pometia pinnata |
| rice | Oryza sativa | tea tree | Melaleuca cajuputi |
| rosewood | Pterocarpus indicus | teak | Tectona grandis |
| rubber tree | Ficus elastica | tiktik | Saccharum spp. |
| sago | Metroxylon sagus | tirip | Cocos nucifera (dry) |
| | M. rumphii | tobacco | Nicotiana tabacum |
| saksak | Metroxylon sagus | tomato | Lycopersicum esculentum |
| | M. rumphii | ton | Pometia pinnata |
| salat | Dendrocnide sp. | tor | Intsia bijuga |
| | Semecarpus sp. | traveller's tree | Ravenala |
| sandalwood | Santalum macgregorii | | madagascariensis |
| sassafras | Dryadodaphne novoguineensis | tree tomato | Cyphomandra betacea |
| sauasap | Annona muricata | tulip | Gnetum gnemon |
| screw pine | Pandanus spp. | utun | Barringtonia speciosa |
| sesame | Sasamum indicum | vanilla | Vanilla fragrans |

| | |
|---|---|
| vut | Derris uglinosa |
| wail limbum | Caryota spp. |
| wail mango | Mangifera spp. |
| walnut | Dracontomelum mangiferum |
| Wau beech | Elmerrillia papuana |
| water cress | Nasturtium officinale |
| water lettuce | Pistia stratiotes |
| watermelon | Citrullus lanatus |
| wattle | Acacia spp. |
| wel | Campnosperma brevipetiolata |
| white beech | Gmelina moluccana |
| white mangrove | Avicennia marina |
| wing bin | Psophocarpus tetragonolobus |
| winged bean | Psophocarpus tetragonolobus |
| yam | Dioscorea spp. |
| yambo | Psidium guajava |
| yar | Casuarina equisetifolia |
| yati | Tectona grandis |
| yellow cheesewood | Sarcocephalus coadunata |
| yellow hardwood | Neonauclea acuminata |
| yerima | Octomeles sumatrana |

# DIVISION PTERIDOPHYTA

### CLASS I.   PSILOPHYTATAE (PSILOTOPSIDA)
**Order 1.   Psilotales**
Family   1. Psilotaceae

### CLASS II.   LYCOPODIATAE (LYCOPSIDA)
**Order 1.   Lycopodiales**
Family   1. Lycopodiaceae
**Order 2.   Selaginellales**
Family   1. Selaginellaceae
**Order 3.   Isoetales**
Family   1. Isoetaceae

### CLASS III.   ARTICULATAE (SPHENOPSIDA)
**Order 1.   Equisetales**
Family   1. Equisetaceae

### CLASS IV.   FILICATAE (FILICOPSIDA)

#### SUBCLASS I.   EUSPORANGIATAE
**Order 1.   Ophioglossales**
Family   1. Ophioglossaceae
**Order 2.   Marattiales**
Family   1. Marattiaceae
2. Osmundaceae
**Order 2.   Plagiogyrales**
Family   1. Plagiogyriaceae

#### SUBCLASS II.   LEPTOSPORANGIATAE
Family   1. Schizaeaceae
2. Parkeriaceae
3. Platyzomataceae
4. Adiantaceae
Subfamily   1. Adiantoideae
2. Vittarioideae
4. Pteridoideae
5. Hymenophyllaceae
6. Gleicheniaceae
7. Matoniaceae
8. Cheiropleuriaceae
9. Dipteridaceae
10. Polypodiaceae
11. Grammitidaceae
12. Cyatheaceae
13. Thyrsopteridaceae

14. Dennstaedtiaceae

Subfamily  1. Dennstaedtioideae

2. Monachosoraceae

3. Lindsaeaceae

15. Thelypteridaceae

16. Aspleniaceae

Subfamily  1. Asplenioideae

2. Athyrioideae

3. Tectarioideae

4. Dryopteridoideae

5. Lomariopsoideae

6. Elaphoglossoideae

17. Davalliaceae

Subfamily  1. Davallioideae

2. Oleandraceae

18. Blechnaceae

**SUBCLASS III.  HYDROPTERIDATAE**

    **Order 1.   Marsileales**

      Family  1. Marsileaceae

    **Order 2.   Salviniales**

      Family  1. Salviniaceae

2. Azollaceae

# DIVISION SPERMATOPHYTA

## SUBDIVISION 1. GYMNOSPERMAE

### CLASS I.   CYCADATAE

    **Order 1.   Cycadales**

      Family  1. Cycadaceae

2. Zamiaceae

### CLASS II.   CONIFERAE

    **Order 1.   Pinales**

      Family  1. Araucariaceae

2. Pinaceae

3. Cupressaceae

    **Order 2.   Taxales**

      Family  1. Podocarpaceae

2. Phyllocladaceae

### CLASS III.  GNETATAE

    **Order 1.   Gnetales**

      Family  1. Gnetaceae

SUBDIVISION 2. ANGIOSPERMAE

CLASS I.  DICOTYLEDONAE

SUBCLASS I.  MAGNOLIIDAE

Order 1.  **Magnoliales**

Family  1. Winteraceae

2. Himantandraceae

3. Eupomatiaceae

4. Austrobaileyaceae

5. Magnoliaceae

6. Annonaceae

7. Myristicaceae

Order 2.  **Laurales**

Family  1. Trimeniaceae

2. Monimiaceae

3. Lauraceae

4. Hernandiaceae

Order 3.  **Piperales**

Family  1. Chloranthaceae

2. Piperaceae

Order 4.  **Aristolochiales**

Family  1. Aristolochiaceae

Order 5.  **Nymphaeales**

Family  1. Nelumbonaceae

2. Nymphaeaceae

3. Cabombaceae

4. Ceratophyllaceae

Order 6.  **Ranunculales**

Family  1. Ranunculaceae

2. Menispermaceae

3. Coriariaceae

4. Sabiaceae

Order 7.  **Papaverales**

Family  1. Papaveraceae

SUBCLASS II. HAMAMELIDAE

Order 1.  **Hamamelidales**

Family  1. Hamamelidaceae

Order 2.  **Daphniphyllales**

Family  1. Daphniphyllaceae

Order 3.  **Urticales**

Family  1. Ulmaceae

2. Cannabidaceae

        3. Moraceae

        4. Cecropiaceae

        5. Urticaceae

**Order 4. Juglandales**

    Family   1. Juglandaceae

**Order 5.  Myricales**

    Family   1. Myricaceae

**Order 6.  Fagales**

    Family   1. Balanopaceae

        2. Fagaceae

**Order 7.  Casuarinales**

    Family   1. Casuarinaceae

# SUBCLASS III.  CARYOPHYLLIDAE

**Order 1.  Caryophyllales**

    Family   1. Phytolaccaceae

        2. Nyctaginaceae

        3. Aizoaceae

        4. Chenopodiaceae

        5. Amaranthaceae

        6. Portulacaceae

        7. Basellaceae

        8. Molluginaceae

        9. Caryophyllaceae

**Order 2.  Polygonales**

    Family   1. Polygonaceae

**Order 3.  Plumbaginales**

    Family   1. Plumbaginaceae

# SUBCLASS IV.  DILLENIIDAE

**Order 1.  Dilleniales**

    Family   1. Dilleniaceae

**Order 2.  Theales**

    Family   1. Ochnaceae

        2. Dipterocarpaceae

        3. Theaceae

        4. Actinidiaceae

        5. Elatinaceae

        6. Clusiaceae

**Order 3.  Malvales**

    Family   1. Elaeocarpaceae

        2. Tiliaceae

        3. Sterculiaceae

        4. Bombacaceae

5. Malvaceae

**Order 4.  Lecythidales**

Family  1. Barringtoniaceae

**Order 5.  Nepenthales**

Family  1. Nepenthaceae

2. Droseraceae

**Order 6.  Violales**

Family  1. Flacourtiaceae

2. Bixaceae

3. Violaceae

4. Ancistrocladaceae

5. Turneraceae

6. Passifloraceae

7. Caricaceae

8. Cucurbitaceae

9. Datiscaceae

10. Begoniaceae

**Order 7.  Capparales**

Family  1. Capparaceae

2. Brassicaceae

3. Moringaceae

**Order 8.  Batales**

Family  1. Bataceae

**Order 9.  Ericales**

Family  1. Clethraceae

2. Epacridaceae

3. Ericaceae

**Order 10.  Ebenales**

Family  1. Sapotaceae

2. Ebenaceae

3. Styracaceae

4. Symplocaceae

**Order 11.  Primulales**

Family  1. Myrsinaceae

2. Primulaceae

**SUBCLASS V. ROSIDAE**

**Order 1.  Rosales**

Family  1. Connaraceae

2. Cunoniaceae

3. Pittosporaceae

4. Byblidaceae

5. Hydrangeaceae

6. Grossulariaceae

7. Alseuosmiaceae

8. Crassulaceae

9. Saxifragaceae

10. Rosaceae

11. Chrysobalanaceae

12. Surianaceae

**Order 2.    Fabales**

Family   1. Caesalpiniaceae

2. Fabaceae

3. Mimosaceae

**Order 3.    Proteales**

Family   1. Elaeagnaceae

2. Proteaceae

**Order 4.    Podostemales**

Family   1. Podostemaceae

**Order 5.    Haloragidales**

Family   1. Haloragidaceae

2. Gunneraceae

**Order 6.    Myrtales**

Family   1. Sonneratiaceae

2. Lythraceae

3. Crypteroniaceae

4. Thymelaeaceae

5. Myrtaceae

6. Punicaceae

7. Onagraceae

8. Melastomataceae

9. Combretaceae

**Order 7.    Rhizophorales**

Family   1. Rhizophoraceae

**Order 8.    Cornales**

Family   1. Alangiaceae

2. Nyssaceae

3. Cornaceae

**Order 9.    Santalales**

Family   1. Olacaceae

2. Opiliaceae

3. Santalaceae

4. Loranthaceae

5. Viscaceae

6. Balanophoraceae

Order 10.   Rafflesiales
    Family   1. Mitrastemmataceae
Order 11.   Celastrales
    Family   1. Celastraceae
             2. Stackhousiaceae
             3. Salvadoraceae
             4. Aquifoliaceae
             5. Icacinaceae
             6. Cardiopteridaceae
             7. Corynocarpaceae
             8. Dichapetalaceae
Order 12.   Euphorbiales
    Family   1. Pandaceae
             2. Euphorbiaceae
             3. Stilaginaceae
Order 13.   Rhamnales
    Family   1. Rhamnaceae
             2. Leeaceae
             3. Vitaceae
Order 14.   Linales
    Family   1. Erythroxylaceae
             2. Ixonanthaceae
             3. Linaceae
Order 15.   Polygalales
    Family   1. Malpighiaceae
             2. Polygalaceae
             3. Xanthophyllaceae
Order 16.   Sapindales
    Family   1. Staphyleaceae
             2. Sapindaceae
             3. Burseraceae
             4. Anacardiaceae
             5. Simaroubaceae
             6. Meliaceae
             7. Rutaceae
             8. Zygophyllaceae
Order 17.   Geraniales
    Family   1. Oxalidaceae
             2. Geraniaceae
             3. Tropaeolaceae
             4. Balsaminaceae
Order 18.   Apiales

Family   1. Araliaceae

            2. Apiaceae

**SUBCLASS VI.  ASTERIDAE**

  **Order 1.  Gentianales**

Family   1. Loganiaceae

            2. Gentianaceae

            3. Apocynaceae

            4. Asclepiadaceae

  **Order 2.  Solanales**

Family   1. Solanaceae

            2. Convolvulaceae

            3. Menyanthaceae

  **Order 3.  Lamiales**

Family   1. Boraginaceae

            2. Verbenaceae

            3. Avicenniaceae

            4. Lamiaceae

  **Order 4.  Callitrichales**

Family   1. Callitrichaceae

  **Order 5.  Plantaginales**

Family   1. Plantaginaceae

  **Order 6.  Scrophulariales**

Family   1. Oleaceae

            2. Scrophulariaceae

            3. Myoporaceae

            4. Orobanchaceae

            5. Gesneriaceae

            6. Acanthaceae

            7. Pedaliaceae

            8. Bignoniaceae

            9. Lentibulariaceae

  **Order 7.  Campanulales**

Family   1. Pentaphragmataceae

            2. Sphenocleaceae

            3. Campanulaceae

            4. Lobeliaceae

            5. Stylidiaceae

            6. Goodeniaceae

  **Order 8.  Rubiales**

Family   1. Rubiaceae

  **Order 9.  Dipsacales**

Family   1. Caprifoliaceae

Order 5.  Cyperales

Family  1. Cyperaceae

  2. Poaceae

Order 6.  Typhales

Family  1. Sparganiaceae

  2. Typhaceae

SUBCLASS IV.  ZINGIBERIDAE

Order 1.  Bromeliales

Family  1. Bromeliaceae

Order 2.  Zingiberales

Family  1. Strelitziaceae

  2. Heliconiaceae

  3. Musaceae

  4. Zingiberaceae

  5. Cannaceae

  6. Marantaceae

SUBCLASS V.  LILIIDAE

Order 1.  Liliales

Family  1. Philydraceae

  2. Pontederiaceae

  3. Haemodoraceae

  4. Iridaceae

  5. Liliaceae

  6. Amaryllidaceae

  7. Agavaceae

  8. Hypoxidaceae

  9. Xanthorrhoeaceae

  10. Hanguanaceae

  11. Taccaceae

  12. Stemonaceae

  13. Smilacaceae

  14. Dracaenaceae

  15. Sansevieraceae

  16. Dioscoreaceae

Order 2.  Orchidales

Family  1. Burmanniaceae

  2. Corsiaceae

  3. Orchidaceae

## Additional References:

Airy Shaw, H.K. (1980): The Euphorbiaceae of New Guinea.- Kew Bulletin Additional Series VIII, 243 pp.

Henty, E.E. (ed., 1981): Handbooks of the Flora of Papua New Guinea Vol. II.- Melbourne University Press, 276 pp.

Johns R.J. (1987-1989): The flowering plants of Papuasia - Dicotyledons.- Parts 1-3, P.N.G. University of Technology.

Johns, R.J., Hay, A.J.M. (eds., 1981-1984): A guide to the monocotyledons of Papua New Guinea.- Parts 1-3, Papua New Guinea

Forestry College (Part 1) and Department of Forests (Parts 2 and 3).

Leach, G.J., Osborne P.L. (1985): Freshwater plants of Papua New Guinea.- The University of Papua New Guinea Press, 254 pp.

Percival, M., Womersley, J.S. (1975): Floristics and ecology of the mangrove vegetation of Papua New Guinea.- Botany Bulletin No. 8, Department of Forests, Division of Botany, 96 pp.

Van Royen, P. (1980): The alpine flora of New Guinea.- Parts 1-3, J. Cramer, Vaduz.

Verdcourt, B. (1979): A manual of New Guinea legumes.- Botany Bulletin No. 11, Office of Forests, Division of Botany, 645 pp.

Womersley, J.S. (ed., 1978): Handbooks of the Flora of Papua New Guinea Vol. I.- Melbourne University Press, 278 pp.

# Publications of Wau Ecology Institute

Menzies, J.I. (1975): Handbook of common New Guinea frogs.- Wau Ecology Institute Handbook No. 1, 75 pp.

Lamb, K.P., Gressitt, J.L. (eds., 1976): Ecology and Conservation in Papua New Guinea.- Wau Ecology Institute Pamphlet No. 2.

Gressitt, J.L., Hornabrook, R.W. (1977, 1985): Handbook of New Guinea beetles.- Wau Ecology Institute Handbook No. 2, 87 pp.

Simon, M. (1977): Guide to biological terms in Melanesian Pidgin.- Wau Ecology Institute Handbook No. 3, 115 pp.

Beehler, B. (1978) Guide to the montane birds of Northeast New Guinea.- Wau Ecology Institute Handbook No. 4, 156 pp.

Gressitt, J.L., Nadkarni, N. (1978): Guide to Mt. Kaindi: Background to montane New Guinea ecology.- Wau Ecology Institute Handbook No. 5, 135 pp.

Menzies, J.I., Dennis, E. (1979): Handbook of New Guinea rodents.- Wau Ecology Institute Handbook No. 6, 68 pp.

McCoy, M. (1980): Reptiles of the Solomon Islands.- Wau Ecology Institute Handbook No. 7, 82 pp.

Hadden, D. (1981): Birds of the North Solomons.- Wau Ecology Institute Handbook No. 8, 109 pp.

Beehler, B., Pratt, T.K., Zimmermann, D.A. (1986): Birds of New Guinea.- Wau Ecology Institute Handbook No. 9, Princeton University Press, 293 pp.

Goeltenboth, F. (ed., 1985, 1990): Subsistence agriculture improvement.- Wau Ecology Institute Handbook No. 10, 232 pp.

Kube, R. (1990): Tropical lowland rain forests.- Wau Ecology Institute, Papua New Guinea Environmental Serries No. 2, 56 pp.

Woodley, E. (ed., 1991): Medicinal plants of Papua New Guinea. Part 1: Morobe Province. Wau Ecology Institute Handbook No. 11, Verlag Joseph Margraf, 158 pp.

Parsons, M. (1992): Butterflies of the Wau-Bulolo valley.- Wau Ecology Institute Handbook No. 12.

www.ingramcontent.com/pod-product-compliance
Lightning Source LLC
Chambersburg PA
CBHW080423270326
41929CB00018B/3144